LAND
APPLICATION OF
SLUDGE

Food Chain Implications

A. L. PAGE
T. J. LOGAN
J. A. RYAN

CRC Press
Taylor & Francis Group
Boca Raton London New York

CRC Press is an imprint of the
Taylor & Francis Group, an **informa** business

First published 1987 by CRC Press
Taylor & Francis Group
6000 Broken Sound Parkway NW, Suite 300
Boca Raton, FL 33487-2742

Reissued 2018 by CRC Press

Library of Congress Cataloging-in-Publication Data

Land application of sludge: Food chain implications

"Proceedings of a workshop . . . sponsored by the
U.S. Environmental Protection Agency, Cincinnati,
Ohio, the University of California at Riverside,
and the Ohio State University . . . [and] conducted
in Las Vegas, Nevada, November 13–15, 1985"—P.
    Bibliography: p.
    Includes index.
    1. Sewage sludge as fertilizer—Congresses.
2. Plants, Effect of trace elements on—Congresses.
3. Trace elements—Congresses.   4. Crops—Composition—
Congresses.   5. Crops and soils—Congresses.   6. Sewage
sludge as fertilizer—Hygienic aspects—Congresses.
7. Health risk assessment—Congresses.   I. Page,
Albert L.   II. United States. Environmental Protection
Agency.   III. University of California, Riverside.
IV. Ohio State University.
S657.E34   1987   631.8'69   87-2925
ISBN 0-87371-083-5

A Library of Congress record exists under LC control number: 87002925

ISBN 13: 978-1-315-89483-6 (hbk)
ISBN 13: 978-1-351-07393-6 (ebk)

Visit the Taylor & Francis Web site at http://www.taylorandfrancis.com and the
CRC Press Web site at http://www.crcpress.com

Preface

The disposal and/or recycling of sewage sludge is a problem facing municipalities throughout the world. As steps are taken to maintain and improve the quality of surface waters, the quantities of sludge generated continue to increase, and municipalities are confronted with an urgent need to develop safe and feasible alternative practices for sludge management. One alternative, as old as sludge treatment itself, and which presumably could accommodate increasing quantities of sludge, is agricultural utilization. The application rates to agricultural lands should be such that the plant nutrients in the sludge (e.g., nitrogen and phosphorus) meet entirely or in part the needs of the crop. However, in addition to valuable plant nutrients, sludges contain a variety of organic and inorganic trace constituents potentially harmful to plants and consumers. As examples, Cu, Ni, Zn, and Cd are harmful to plants at relatively low concentrations, particularly plants grown on acid soils. Cadmium is of even greater concern because of its harmful effects on plants, animals, and man. Toxic trace organics (e.g., polychlorobiphenyls, or PCBs) likewise may find their way into crops and subsequently through the food chain to humans. Consequently, disposal/recycling of municipal sewage sludge on agricultural lands must be carried out in a way which provides a beneficial use to the crop without developing pollution problems associated with the accumulation of trace constituents in soils.

A difficulty facing regulatory agencies and municipalities interested in adopting the option of applying municipal sludge to agricultural lands is one of knowing the maximum annual permissible application rate as well as long-term rates which will not result in harmful accumulation of trace constituents in soils and crops. Recent information now entering or about to enter the scientific literature suggests that the approach used to estimate dietary intakes arising from sludge use on land could be improved. Likewise, recent studies suggest that certain factors now known to affect trace constituent uptake by plants were not considered in previously developed guidelines and criteria for land application of sludge. For example, although recent studies show that bioavailability of trace constituents in sludges is related to their overall composition and type of treatment—e.g., sludges with or without $FeCl_3$ or $Al_2(SO_4)_3$ treatment—present day risk assessment scenarios assume that bioavailability is independent of sludge quality and type.

iii

Recognizing the need to utilize the most current available information in the development of regulations and criteria to safely manage land application of municipal sludge, the U.S. Environmental Protection Agency, in cooperation with the University of California, Riverside, and the Ohio State University, Columbus, sponsored a workshop on Effects of Sewage Sludge Quality and Soil Properties on Plant Uptake of Sludge-Applied Trace Constituents. The workshop brought together 31 scientists knowledgeable in the subject matter to critically examine current available published and unpublished information and produce in report form an assessment of the current knowledge about factors known to affect the impact of trace constituents on crops and consumers when applied to lands in the form of municipal sludge. This book presents the findings of the workshop.

Albert L. Page
Terry J. Logan
James A. Ryan

Acknowledgments

This book is the proceedings of a workshop entitled *Effects of Sewage Sludge Quality and Soil Properties on Plant Uptake of Sludge-Applied Trace Constituents*, jointly sponsored by the U.S. Environmental Protection Agency, Cincinnati, OH, the University of California at Riverside, and The Ohio State University in Columbus. The meeting was conducted in Las Vegas, NV, November 13–15, 1985, under cooperative agreement CR-812673, and the project officer, representing the Cincinnati EPA's Wastewater Engineering Research Laboratory, was J. A. Ryan. This book has been reviewed by the U.S. Environmental Protection Agency and, though it is approved for publication, mention of trade names or commercial products does not constitute endorsement or recommendation for use by either the EPA or by the Publisher.

Individuals from a number of federal agencies, colleges, universities, and municipalities participated in the workshop and contributed to its findings. We thank them for their contributions and their institutions for allowing them to participate. Names and affiliations of all workshop participants are presented in the concluding section of the book. The assistance of Dr. Alan Rubin, OWRS, USEPA, Washington, DC, and Mr. Randall Bruins and Mr. Larry Fradkin, ECOA, USEPA, Cincinnati, OH, in the organization and planning of the workshop is gratefully acknowledged.

Special thanks are due to those who chaired workgroup sessions and assumed the leadership responsibility for preparation of reports. Lee Sommers, Colorado State University, Fort Collins, CO, and V. Van Volk, Oregon State University, Corvallis, OR, chaired the workgroup on Effects of Soil Properties on Accumulation of Trace Elements by Crops; Richard Corey, University of Wisconsin, Madison, WI, chaired the workgroup on Effects of Sludge Properties on Accumulation of Trace Elements by Crops; Andrew Chang, University of California, Riverside, CA, chaired the workgroup on Effects of Long-Term Sludge Application on Accumulation of Trace Elements by Crops; Rufus Chaney, USDA-ARS-NER, Beltsville, MD, chaired the workgroup on Transfer of Sludge-Applied Trace Elements to the Food Chain; and Lee W. Jacobs, Michigan State University, East Lansing, MI, chaired the workgroup on Effects of Trace Organics in Sewage Sludge on Soil Plant Systems and Assessing Their Risk to Humans.

We owe a special debt of gratitude to Ms. Louise DeHayes who typed and copyedited camera-ready copy and to Ms. Leslie Webster who assisted in local arrangements and in the typing of manuscripts.

Contents

LAND
APPLICATION OF
SLUDGE
Food Chain Implications

Albert L. Page
University of California, Riverside, California

Terry J. Logan
The Ohio State University, Columbus, Ohio

James A. Ryan
U.S. Environmental Protection Agency, Cincinnati, Ohio

In the last two decades, this nation has experienced a dramatic increase in the construction of publicly-owned treatment works (POTW) with a corresponding increase in residual solids from treating the waste water. Because the common methods of sludge disposal, such as landfill, incineration and ocean dumping may not be adequate or suitable to accommodate the ever-increasing quantities of POTW sludge, interest in applying sludges to agricultural, forest and disturbed land has increased.

In addition to valuable plant nutrients, sewage sludge contains variable concentrations of trace elements and synthetic organic compounds. Concern for trace element contamination of the food chain from land application of sewage sludge stems from extensive prior experience with phytotoxicity of elements such as Cu, Ni and Zn from smelters and other sources (Page, 1974) and from human and livestock toxicities associated with environmental contamination by Pb, Hg, Cd, Cu, F, Mo, As and other trace elements (Logan and Chaney, 1983).

The Joint Conference on Recycling Municipal Sludges and Effluents on Land (1973) raised the issue of trace element contamination from sewage sludge but the available data base on actual land application research was for the most part limited to pot studies with metal salts or sludge and a few field experiments of no more than a few years duration (Logan and Chaney, 1983). Subsequent conferences in 1980 (Council for Agricultural Science and Technology [CAST], 1980) and 1983 (Page et al., 1983) reexamined these issues in light of the increasing body of research. By 1983 Logan and Chaney had concluded that the environmental threat from sludges applied to land at agronomic rates was minimal when existing federal

regulations and guidelines (EPA, 1979) were followed. Phytotoxicity from sludge-applied metals was no longer believed to
be of concern except for high-metal content sludges applied at
high loading rates on acid soils. ·

Inputs of sludge-borne trace elements to agricutural land
in the U.S. has been governed since 1979 by EPA regulations and
guidelines (EPA, 1979). Under various provisions of existing
federal statutes, cadmium was the only trace element addressed
and the regulatory approach was to limit annual and cumulative
applications of Cd to land, based on soil pH and soil cation
exchange capacity (CEC). In addition to federal regulations,
many states imposed limitations on cumulative applications of
elements such as Cu, Ni, Zn (to protect against phytotoxicity)
and Pb (to protect the human food chain) (Logan and Chaney,
1983).

Implicit in this regulatory approach was the belief that
bioavailability of sludge-applied trace elements was controlled
by soil processes such as adsorption, chelation, and precipitation and that these processes were reflected by soil properties
such as pH and CEC. By 1980, however, data from the increasing
number of long-term field studies were beginning to indicate
that sludge properties could also influence trace element bioavailability. The sludge's effect on bioavailability of trace
elements was postulated by Corey et al. (1981) and later reiter-
ated by Logan and Chaney (1983) as being due to binding of trace
elements by the sludge itself. A corollary to this hypothesis is
the prediction that, at high enough sludge application rates, the
solubility of trace elements in soil would be controlled by the
sludge and not by the soil. The implication of this theory, if
true, is that the present regulation of Cd application to land
with sludge on the basis of Cd loading and soil properties alone
ignores what may be the equally or more important sludge properties, and may overestimate crop Cd uptake particularly from
low-Cd sludges.

Parallel to, but more recent than, the evolution of our
knowledge of trace element chemistry and bioavailability is the
growing concern over contamination of the environment by synthetic organic compounds. This concern has led to recent but
limited studies of synthetic organic compounds in sewage sludges
(Naylor and Loehr, 1982a,b; Overcash, 1983; Overcash et al., 1986)
and proposals for their regulation. Under the 1979 regulations,
only polychlorinated biphenyls (PCBs) were specifically controlled (EPA, 1979). The research data base on the fate of
sludge-borne organics is extremely limited, as is the information
on the content of various synthetic organic compounds in sludges.
As a result, uncertainties as to the health effects and threshold
exposures of any of these compounds has made the evaluation of
risk from sludge organics difficult.

In 1984, EPA began a process to reevaluate the existing
regulations and criteria by which land application of municipal
sewage sludge is controlled in the U.S. The Office of Water
Regulations and Standards working with several technical advisory
committees and with the Environmental Criteria and Assessment
Office (ECAO) screened those pollutants found in sludge that had

the potential to adversely affect the food chain, thus possibly requiring regulation. Based on this evaluation, and using a risk assessment approach developed by ECAO, hazard indices were developed for a number of trace elements and synthetic organic compounds, and subsequently were used to evaluate the potential risk from land application of sludge (EPA, 1985). Presently, a comprehensive risk assessment methodology which will be used to evaluate potential risk from land application of municipal sludge is under development by ECAO Cincinnati. The result of this and related efforts will be development of revised or, if necessary, new regulations governing land application and other means of sludge use and disposal.

The development of hazard indices and their use in risk assessment is limited by the availability of valid data for the pollutants of concern. A critical review of the data bases used to develop the hazard indices revealed that they often included studies involving metal salts addition rather than sludge only sources and did not include many of the long-term field studies with sludges which were conducted in the late 1970's and early 1980's and which are now beginning to enter the literature.

Although the complete list of trace elements for which hazard indices were developed was addressed in the workshop, the focus of Chapters 2-6 was limited to Cd, Zn, Mo, Fe, Pb and Se, as the other trace elements usually present in sludges were considered to present little potential risk to the human or animal food chain.

REFERENCES

Council for Agricultural Science and Technology (CAST). 1980. Effects of sewage sludge on the cadmium and zinc content of crops. Council for Agricultural Science and Technology Report No. 83. Ames, IA.

Corey, R. B., R. Fujii, and L. L. Hendrickson. 1981. Bio-availability of heavy metals in soil-sludge systems. p. 449-465. In Proc. Fourth Annual Madison Conf. Appl. Res. Pract. Municipal Ind. Waste. Madison, WI, 28-30 Sept. 1981. Univ. of Wisconsin-Extension, Madison, WI.

Environmental Protection Agency (EPA). 1979. Criteria for classification of solid waste disposal facilities and practices; final interim final, and proposed regulations. (40 CFR 257). Fed. Reg. 44(179):53460-53468.

Environmental Protection Agency. 1985. Summary of environmental profiles and hazard indices for constituents of municipal sludge. Office of Water Regulations and Standards, Washington, D.C. 20460. 64 p.

Joint Conference on Recycling Municipal Sludges and Effluents on Land. 1973. Natl. Assn. State Univ. and Land Grant Colleges, Dupont Circle, Washington, D.C. 20036.

Logan, T. J. and R. L. Chaney. 1983. Utilization of municipal wastewater and sludge on land—Metals. p. 235–326. In A. L. Page et al. (ed.) Proc. of the workshop on utilization of municipal wastewater and sludge on land. University of California, Riverside, CA.

Naylor, L. M., and R. C. Loehr. 1982a. Priority pollutants in municipal sewage sludge. Biocycle, July/August, p. 18-22.

Naylor, L. M., and R. C. Loehr. 1982b. Priority pollutants in municipal sewage sludge. Part II. Biocycle, November/December, p. 37-42.

Overcash, M. R. 1983. Land treatment of municipal effluent and sludge: Specific organic compounds. p. 199–231. In A. L. Page, et al. (ed.) Proc. of the workshop on utilization of municipal wastewater and sludge on land. Univ. of California, Riverside, CA.

Overcash, M. R., J. R. Weber, and W. P. Tucker. 1986. Toxic and priority organics in municipal sludge land treatment systems. EPA 600/2-86-010, U.S. EPA, Cincinnati, OH.

Page, A. L. 1974. Fate and effects of trace elements in sewage sludge when applied to agricultural lands. A literature review-study. U.S. EPA Report No. EPA-670/2-74-005. National Environmental Research Center, Environmental Protection Agency, Cincinnati, OH.

Page, A. L., T. L. Gleason III, J. E. Smith Jr., I. K. Iskandar, and L. E. Sommers (ed.) 1983. Proc. of the workshop on utilization of municipal wastewater and sludge on land. Univ. of California, Riverside, CA.

CHAPTER 2

EFFECTS OF SOIL PROPERTIES ON ACCUMULATION OF TRACE ELEMENTS BY CROPS

Lee Sommers
Colorado State University, Fort Collins, Colorado

V. Van Volk
Oregon State University, Corvallis, Oregon

Paul M. Giordano
Tennessee Valley Authority, Muscle Shoals, Alabama

William E. Sopper
Pennsylvania State University, University Park, Pennsylvania

Robert Bastian
U. S. Environmental Protection Agency, Washington, DC

INTRODUCTION

The fate and effects of sewage sludge constituents in a soil-plant system are influenced by factors such as climate (rainfall and temperature), management (irrigation, drainage, liming, fertilization, addition of amendments), and composition of the sewage sludge. In addition, soil properties affect the chemical reactions and processes which occur after application of sewage sludge to a soil. Soil properties that affect the reactions and resultant plant uptake of sewage sludge constituents include pH, organic matter, cation exchange capacity, iron and aluminum oxides, texture, aeration, specific sorption sites and water availability. Mean values for selected soil properties are shown in Table 1. Many of these factors are interrelated and thus create a rather complex medium involving chemical and microbial reactions. The factors which tend to be stable are texture, CEC, organic matter, and iron and aluminum oxides. Factors such as pH, water content, and aeration (relates to water content) vary frequently or are easier to adjust. For example, soil pH can be increased by lime additions while ammoniacal fertilizers acidify soils.

Table 1. Metal and organic carbon contents, CEC, and pH for soils from selected sites in the continental United States (from Holmgren et al., 1987).*

State	Nt	Cd Mean	SD‡	Pb Mean	SD	Zn Mean	SD	Cu Mean	SD	Ni Mean	SD	pH Mean	SD	Organic C Mean	SD	CEC Mean	SD
		(mg/kg)												(%)		(mmol(+)/kg)	
Arizona	14	0.24	0.06	14	4	72	18	39.4	11.3	28.7	6.5	7.7	0.5	0.37	0.14	143	36
Iowa	85	0.24	0.07	14	4	62	16	21.3	5.7	28.2	8.1	5.9	0.7	2.53	0.94	282	77
Missouri	33	0.27	0.08	20	5	60	9	18.8	3.8	24.8	3.5	6.5	0.7	1.80	0.61	200	46
Minnesota	89	0.30	0.09	12	2	71	20	22.3	4.5	30.0	5.5	5.9	0.8	3.01	0.90	342	89
California	283	0.31	0.28	12	12	93	41	46.6	30.2	74.3	63.9	7.2	0.8	1.00	0.88	214	140
Kansas	38	0.32	0.07	15	2	53	10	15.6	2.5	20.4	3.4	5.6	0.9	1.14	0.18	192	25
W.Virginia	40	0.32	0.34	646	1127	84	37	96.9	143	23.3	13.1	5.3	0.7	2.99	2.19	141	78
Wisconsin	164	0.35	0.20	12	7	59	29	37.7	36.8	16.4	9.4	5.9	0.8	14.64	16.45	556	552
Montana	33	0.37	0.07	11	2	75	13	21.0	4.6	26.1	5.5	6.8	0.9	1.41	0.38	171	35
N.Dakota	30	0.37	0.21	10	5	69	42	22.0	12.9	31.1	16.9	7.1	0.8	1.99	0.76	260	151
Idaho	54	0.38	0.16	11	2	68	24	22.0	8.0	25.2	7.9	7.4	1.0	1.16	0.54	173	39
Ohio	81	0.38	0.15	19	4	89	41	28.1	11.7	28.2	9.7	6.4	0.6	1.83	0.54	189	58
Colorado	89	0.39	0.30	16	13	85	42	19.4	8.0	15.9	7.2	7.7	0.5	0.80	0.29	137	49
Nebraska	72	0.39	0.18	14	3	58	25	17.3	7.7	21.7	12.3	6.4	0.8	1.49	0.48	199	51
Florida	89	0.44	0.28	10	10	88	67	103.7	88.2	10.3	7.7	5.8	0.9	26.80	18.11	970	657
New York	173	0.45	0.36	17	5	64	37	74.8	77.5	19.5	10.1	5.4	0.8	16.71	17.32	767	769
Oregon	106	0.49	0.45	11	7	71	30	33.4	16.5	27.1	6.0	6.1	1.0	3.36	5.03	338	279
S.Dakota	44	0.56	0.12	14	3	96	25	30.3	8.8	42.3	18.8	6.5	0.9	2.62	0.49	300	49
Michigan	86	0.94	0.30	16	6	80	37	111.5	75.6	14.7	8.6	5.7	0.7	28.43	5.13	1358	315
S.Carolina	10	0.03	0.01	10	3	12	5	4.1	1.7	4.1	1.8	4.2	0.3	2.27	1.00	81	22
Georgia	146	0.05	0.05	8	5	18	19	7.0	5.3	9.0	7.9	5.9	0.5	0.74	0.26	35	15
Alabama	92	0.06	0.07	7	4	16	7	8.1	5.9	11.3	6.8	5.8	0.7	0.65	0.30	31	19
Maryland	57	0.08	0.02	11	4	31	16	8.1	2.6	12.4	4.4	5.7	0.7	0.75	0.17	44	21
N.Carolina	242	0.09	0.07	10	6	15	14	8.9	12.9	8.6	13.1	5.1	0.5	0.69	1.74	74	84
Oklahoma	94	0.10	0.08	7	3	33	47	4.0	13.3	14.9	14.4	6.4	0.6	0.65	0.23	94	54
Virginia	46	0.14	0.07	98	118	59	29	9.4	28.4	22.3	12.0	5.6	0.8	2.07	0.97	92	34
Texas	362	0.16	0.11	9	5	40	27	2.2	6.9	15.9	10.2	7.1	1.0	0.78	0.41	153	105
Delaware	4	0.17	0.06	10	2	25	9	5.0	2.2	6.6	4.4	6.3	1.1	0.55	0.22	4	1
Maine	31	0.17	0.03	13	2	74	13	0.7	24.0	41.5	6.4	4.5	0.5	2.25	0.43	134	18
Arkansas	62	0.18	0.15	15	8	45	33	5.5	8.8	17.2	9.8	5.7	0.3	1.07	0.32	145	97
Illinois	135	0.20	0.09	16	3	56	21	7.2	5.6	20.6	7.0	6.0	0.9	1.79	0.77	195	82
New Mexico	41	0.20	0.07	11	3	47	14	6.1	5.7	16.9	5.0	8.2	0.7	0.58	0.19	142	48
Washington	122	0.20	0.08	9	4	66	19	7.3	10.5	29.0	16.7	6.3	0.1	1.48	2.71	141	59
Pennsylvania	45	0.21	0.24	24	25	30	27	5.3	28.4	10.4	7.8	6.1	0.9	1.36	0.59	90	24
Louisiana	133	0.22	0.14	16	16	64	51	2.1	16.6	25.3	18.7	5.7	0.6	1.37	0.65	238	174
Indiana	80	0.23	0.14	13	5	51	25	7.0	9.8	16.5	8.6	5.7	0.7	1.35	0.55	130	57

*Soils were selected from sites removed from mobile and point source contamination; values reported for metals may not be representative of U.S. soils in general. Data are expressed on a dry weight basis.
†N= no of soil sites analyzed   ‡SD = standard deviation

Soil cation exchange capacity (CEC) is dependent on soil properties such as organic matter, pH, and type and percentage of clay. Thus it serves as an easily measured, integrating parameter to characterize a soil. Soil pH, like CEC, is an easily measured soil property which provides background information relevant to assessing elemental availability to plants. The soil pH measured in the laboratory is a representation of that which occurs under field conditions. The pH at any individual site in the soil may be significantly different from the pH of other sites. For example, the pH at the root-soil interface may be lower because of exuded organic acids. Due to differential uptake of cations and anions, the pH in the root cylinder of active root hairs may be lower than in older parts of the root system (Römheld and Marschner, 1986). Also, pH reductions with time in sludge-treated soils are due to protons generated during oxidation of reduced forms of N and S mineralized from sludge organic matter. Similar pH reductions occur after addition of fertilizers, particularly those containing ammonium.

Plant uptake of elements from the soil solution initially requires positional availability to the plant root. Either the element must be moved to the root through diffusion or mass flow processes, or the root must grow to the element. The element must then occur in a form which can move into the plant via the uptake mechanism. This transfer requires that the element move through a solution phase; thus, water solubility and a variety of complexation, chelation, and other-chemical reactions become important.

Considerable research on microelement uptake by plants has been done with metal salts. However, metals applied to a soil as a salt, commonly a sulfate, chloride, or nitrate salt, accumulate in plants more readily than the same quantity of metal added in sewage sludge (Logan and Chaney, 1983; Dijkshoorn et al., 1981). Metal salt additions to soils can cause formation of metal chloride complexes and ion pairs which may increase metal diffusion and plant uptake (Bingham, 1980). Metals in sludges are often associated with the insoluble inorganic components (such as phosphates, sulfides, and carbonates) and are not readily plant available (Soon, 1981; Page, 1974). Elemental uptake by plants grown in soils treated with metal salts or sewage sludge amended with metal salts will be higher than actually exists for equal amounts of metal contained in sewage sludges. If results from sludge-treated soils are available, human or animal exposure models should be based on these observations and not on extrapolation of data from additions of soluble metal salts to soils.

To predict the impact of sludge use on elemental content in the human diet, plant uptake of trace elements from sewage sludge should be measured in field experiments. Greenhouse or pot study experiments generally create a root environment which increases the magnitude of trace element uptake (deVries and Tiller, 1978; Davis, 1981). The enhanced uptake of trace elements generally results from four factors: 1) the use of acid forming fertilizers; 2) increased soluble salt content from fertilizers in a smaller soil volume than in the field; 3) root confinement; and 4) unnatural watering patterns. However, greenhouse pot experiments can have value if plants are harvested in an early growth stage or if pots are sufficiently large to allow unrestricted root growth and natural water drainage. In addition, pot experiments are valuable for evaluating factors affecting plant uptake of trace elements, realizing that plant concentrations may differ from those found in a field study.

BACKGROUND LEVELS OF TRACE ELEMENTS

The trace element content of crops is a function of the plant available level in the soil and the modifying influences of soil chemical and physical properties. Trace element levels of soil vary with the parent material. Except for a few special cases (Lund et al., 1981), plant tissue concentrations are not positively correlated with the total trace element content in untreated soils.

Background levels of metals have been summarized for soils from Ohio (Logan and Miller, 1983), Minnesota (Pierce et al., 1982), and from 3,305 sites across the U.S. (Table 1; Holmgren et al., 1987). Although quite extensive, the data contained in Table 1 were selected on the basis of being agricultural soils removed from mobile and point sources of contamination and means representative of all agricultural soils in the U.S. may differ from those presented in Table 1. For most elements, the minimum and maximum values differ by 2 to 3 orders of magnitude. The

unusually high mean values for Pb in soils from Virginia and West
Virginia are due to metalliferous deposits near a few sites. One
statistical approach to evaluating whether a soil has been
impacted by industrial sources of metals is to compare trace
element concentrations with the concentrations at the 95th percentile. As shown, the 95th percentile for soil metals (Table 2)

Table 2. Trace element concentrations for soils from selected sites in
the continental United States (Holmgren et al., 1987).[*]

	Cd	Pb	Zn	Cu	Ni
	- - - - - - - - - - - - - mg/kg - - - - - - - - - - - -				
Mean	0.27	17	57	30	24
Median	0.20	11	54	19	18
Geometric mean	0.17	11(16)[*]	43(48)	18(17)	16(13)
Std. deviation	0.26	141	39	42	27
Maximum	2.3	4,109	402	735	269
Minimum	0.01	0.2	1.5	0.3	0.3
50th percent.	0.20	11	54	19	18
95th percent.	0.79	26	127	98	56

[*]Soils were selected from sites removed from mobile and point source
contamination; values reported for metals may not be representative of
U.S. soils in general. Data are expressed on a dry weight basis.

[*]Geometric mean of U.S. soils from Shacklette and Boerngen (1984).

is appreciably smaller than the maximum value. Current U.S.
EPA Cd limits imposed on sludge applications (EPA, 1979) are 5
to 10 fold greater than a liberal estimate of natural background
levels. Further, a soil's total trace element content enables
a preliminary evaluation of metal contamination from prior waste
disposal activities. The total soil metal data also allows identification of sites where parent material contains unusually high
levels of a given element.
    The total metal concentration reported for soils may be
influenced by the analytical methods employed, especially if a
dissolution procedure is used. The total metal content in a
soil requires either a non-destructive analytical method such
as neutron activation analysis or a total dissolution of the
soil matrix with strong acids plus HF, or partial dissolution
with boiling 4 $\underline{M}$ HNO_3, or refluxing HNO_3-HClO_4 (Lund et al.,
1981). Once the soil matrix is dissolved, standard atomic
absorption or equivalent methods can be used to analyze metals
in the digest.
    A need still exists for a standard extractant to assess the
level of plant available metals in soils. Logan and Chaney
(1983) summarize recent research on common soil metal extractants. The DTPA-TEA reagent used to detect trace metal deficiencies in calcareous soils (Lindsay and Norvell, 1978) has
been used to monitor soils. Other extractants which have been
used include double acid (HCl + H_2SO_4), dilute HCl, $Ca(NO_3)_2$,
and water saturation extracts. One approach is the diagnostic
soil test used on a routine basis for soils treated with sludges
in Pennsylvania (Baker and Amacher, 1981). This method involves
equilibrating soils with a test solution containing cations (Ca,
Mg, K, and H) at the activities and ratios determined to be near
the minimum for optimum growth of plants. The solution also
contains $4x10^{-4}$ $\underline{M}$ DTPA (diethylene triamine pentaacetic acid) to

render a small exchange of trace metals. The extracted metal
provides a measure of the labile pool and the metal-DTPA formation constants are used to calculate activities of trace metals
(Baker and Amacher, 1981). However, no method used has been
proven acceptable to predict plant uptake of metals from a wide
range of soils (Logan and Chaney, 1983). Ideally, extraction of
a soil, sewage sludge, or a sludge amended soil could be used to
predict the eventual plant uptake of trace elements. With this
approach, bioavailable elements of both the soil and sewage
sludge could be assessed. A procedure is also needed for quan-
titative measurement of specific sorption sites in soils.

Information on the background levels of metals in crops is
also needed to evaluate the impact of metals entering animal or
human diets. A recent survey has been conducted by the USDA-EPA-
FDA for a variety of crops grown in major regions of the U.S.
(Wolnick et al., 1983a,b, 1985). As previously discussed for
soils from the same survey, the data presented in Table 3 were

Table 3. Trace element concentrations (dry weight) in the edible part of crops grown on untreated soils.[a]

Crop	Cd[b]				Zn[c]				Pb[b]			
	Min	Max	Median	95th[d]	Min	Max	Median	95th	Min	Max	Median	95th
	- - - - µg/kg - - - -				- - - mg/kg - - - -				-------------µg/kg---------			
Lettuce	34	3800	435	2100	13	110	46	78	36	1700	190	994
Spinach	160	1900	800	1480	17	200	43	128	240	2300	530	1180
Potatoes	9	1000	140	360	5.1	35	15	27	1	2200	25	97
Wheat	5	220	36	125	11	76	29	48	1	770	21	168
Rice	<1	250	5	34	7.7	23	15	20	<1	80	5	26
Sweet corn	0.5	230	8	57	28	55	25	46	7.6	260	9	62
Field corn	<1	350	4	67	12	39	22	30	<1	3600	6	32
Carrots	15	1200	160	786	3.8	61	20	48	10	1100	55	236
Onions	11	340	90	240	6.1	33	16	26	2	720	38	95
Tomatoes	45	790	220	610	12	35	22	29	<1	460	27	108
Peanuts	11	660	68	219	17	63	31	42	<1	200	8	27
Soybeans	1	1200	45	180	32	70	45	59	3	350	36	99

[a]Data are for crops grown in areas removed from mobile and point sources of contamination.
[b]Wolnick et al. (1983a,1983b)
[c]Wolnick et al. (1985)
[d]95th percentile

for crops grown on soils removed from mobile and point sources
of contamination. The data, therefore, may not be representative of the crops for U.S. agricultural soils in general. The
Cd, Zn and Pb content of 12 common crops varies by 1 to 3 orders
of magnitude (Table 3). Median concentrations of Cd in leafy
vegetables were highest (spinach, 800 µg/kg); median concentrations of Cd in root crops ranged from 68 µg/kg (peanuts) to 160
µg/kg (carrots); and for grains the median Cd concentrations
varied from 4 µg/kg (field corn) to 45 µg/kg (soybeans). Somewhat similar median concentrations of Pb in crops were observed
(Table 3). Median concentrations of Zn across the 12 crops
tested, however, were more uniform. They varied from a low of
15 mg/kg (rice) to a high of 46 mg/kg (lettuce).

REGIONAL STUDY OF SLUDGE USE

The regional project W-124 (Optimum utilization of Sewage
Sludge on Cropland) has collected data on the uptake of metals

Table 4. General characteristics of soils used in W-124 study.

Location	Soil	Sand	Silt	Clay	Cation Exch. capacity	Organic C
		--- g/kg ---			(mmol(+)/kg)	(g/kg)
AL	Decatur clay (Rhodic Paleudult)	240	340	420	91	6.6
AZ	Pima clay loam (Typic Torrifluvent)	330	340	300	280	11.0
CA(D)	Domino loam (Xerollic Calciorthid)	200	280	420	140	8.0
CA(G)	Greenfield sandy loam (Typic Haploxeralf)	100	270	630	90	5.0
CO	Nocono clay loam (Aridic Argiustoll)	250	400	350	230	13.2
FL	Lake fine sand (Typic Quartzipsamment)	942	35	23	14	7.0
IL	Ipava silt loam (Aquic Argiudoll)	-	-	-	266	35.1
IN	Chalmers silt loam (Typic Haplaquoll)	120	610	270	251	24.0
MD*	Christiana sandy loam (Typic Paleudult)	-	-	-	59	10.5
MI	Celina silt loam (Aquic Hapludalf)	510	380	110	120	15.0
MN	Port Byron silt loam (Typic Hapludoll)	90	660	250	220	22.5
NE	Sharpsburg silty clay (Typic Argiudoll)	50	610	350	23	20.3
OH	Celina silt loam (Aquic Hapludalf)	230	610	160	120	18.0
OR	Willamette silt loam (Pachic Ultic Argixeroll)	360	766	198	150	22.0
UT	Millville silt loam (Typic Haploxeroll)	340	530	130	190	11.6
WI	Plano silt loam (Typic Argiudoll)	36	820	140	197	23.0

*MD(L) and MD(H) refers to unamended and CaCO₃ amended soils, respectively.

by barley grown at 15 locations in the U.S. (Table 4). At each location, the same sewage sludge sample from Chicago was applied either at 100 mt/ha in the initial year or at 20 mt/ha each year for 5 consecutive years. A 100 mt/ha application of this sludge resulted in addition of 20 kg Cd/ha, the upper limit allowed by current federal regulations (EPA, 1979). Barley was also grown on soils fertilized according to soil test recommendations. Barley leaf and grain and soil samples were collected and analyzed for Cd, Zn, Cu and Ni each year (Tables 5 to 8).

The major conclusions from this experiment are summarized as:

(1) The metal content of barley grain and tissue were similar for untreated and NPK fertilized soils.

(2) Yearly variations in plant metal composition were observed for sludge-treated and untreated plots at most locations.

(3) Metal levels in a plant grown on untreated soil could exceed those found at another location where sewage sludge was added to the soil.

(4) Cadmium concentrations in barley leaf tissue were greater than those in grain.

(5) Sewage sludge application increased metal concentrations for soil and plant tissues with a single 100 mt/ha or annual applications of 20 mt/ha.

(6) The initial application of 20 mt/ha caused a greater increase in plant metal levels than the subsequent 4 annual sludge treatments.

(7) The increase in Zn and Cd in plants grown on sewage sludge treated soils was greater than for Pb, Cu, and Ni.

(8) After the fifth year, the concentrations of Cd and Zn in the barley tissue depended only on the total amount of sludge applied and not upon the frequency of application (i.e., 100 mt/ha in year 1 vs. 20 mt/ha in years 1 through 5).

Table 5. Concentrations of Cd in DTPA soil extracts and in leaf and grain of barley grown in 15 locations.

Location*	Year	DTPA Extract			Barley Leaf			Barley Grain		
		20 mt/ha/y	100 mt/ha	NPK	20 mt/ha/y	100 mt/ha	NPK	20 mt/ha/y	100 mt/ha	NPK
						--mg/kg--					
AZ	1	1.64	4.86	0.30	0.61	1.19	0.46	0.21	0.32	0.03
	2	1.59	4.00	0.03	2.88	3.46	0.83	0.05	0.11	0.03
	3	3.00	4.03	0.25	1.43	2.47	1.95	0.04	0.06	0.05
	4	3.41	3.52	0.25	1.63	2.13	2.00	0.03	0.03	0.03
	5	3.52	3.63	0.19	ND†	ND	ND	ND	ND	ND
CA(G)	1	0.83	3.85	0.10	0.07	0.08	0.03	0.08	0.16	0.03
	2	0.95	1.35	0.10	0.03	0.03	0.03	0.03	0.04	0.03
	3	1.45	1.40	0.10	0.03	0.03	0.03	0.05	0.04	0.04
	4	1.52	1.26	0.10	0.03	0.04	0.03	0.09	0.04	0.06
	5	2.39	1.18	0.06	0.13	0.24	0.06	0.05	0.03	0.05
CA(D)	1	0.75	4.80	0.10	0.03	0.07	0.03	0.03	0.06	0.03
	2	1.48	3.05	0.10	0.03	0.03	0.03	0.03	0.03	0.03
	3	2.05	2.40	0.10	0.03	0.03	0.03	0.04	0.03	0.03
	4	1.95	1.77	0.10	0.03	0.03	0.03	0.03	0.03	0.03
	5	2.13	1.40	0.06	0.04	0.33	0.05	0.09	0.08	0.03
CO	1	ND	ND	ND	ND	ND	ND	0.08	0.16	0.05
	2	0.34	1.65	0.09	0.33	0.55	0.25	0.16	0.22	0.12
	3	0.50	2.05	0.13	ND	ND	ND	0.11	0.33	0.04
	4	1.77	2.29	0.12	ND	ND	ND	0.27	0.73	0.03
	5	1.25	2.63	0.21	ND	ND	ND	0.03	0.03	0.03
FL	1	0.76	4.58	0.06	0.44	0.67	0.33	0.11	0.13	0.03
	2	2.06	4.50	0.07	0.37	0.45	0.30	0.25	0.38	0.06
	3	2.93	4.05	0.05	0.69	0.73	0.20	0.32	0.47	0.03
	4	4.35	5.45	0.15	ND	ND	ND	0.27	0.37	0.05
	5	2.70	3.70	0.17	0.15	0.19	0.07	0.66	0.85	0.16
IN	1	ND	0.39	0.51	ND	ND	ND	ND	0.36	0.07
	2	ND	0.41	0.16	ND	ND	ND	ND	0.86	0.31
	3	ND	2.64	0.12	ND	ND	ND	ND	0.50	0.15
	4	ND	2.41	0.14	ND	0.43	0.33	ND	0.25	0.13
	5	ND	5.35	0.24	ND	0.22	0.08	ND	0.19	0.04
MD(L)	1	ND	ND	ND	0.86	1.22	0.13	0.27	0.78	0.07
	2	0.70	3.75	0.11	0.41	1.22	0.36	0.21	0.54	0.12
	3	0.80	3.04	0.09	0.22	0.43	0.11	0.30	0.32	0.11
	4	2.13	3.20	0.06	0.80	2.65	0.28	0.54	0.43	0.07
	5	3.54	3.13	0.06	0.56	0.79	0.35	0.32	0.42	0.09
MD(H)	1	ND	ND	ND	0.49	1.21	0.15	0.25	0.59	0.06
	2	0.91	4.09	0.06	0.46	0.92	0.18	0.11	0.33	0.03
	3	0.83	2.57	0.05	0.58	0.17	0.07	ND	0.26	0.20
	4	2.28	3.08	0.07	0.60	0.68	0.18	0.40	0.46	0.07
	5	4.09	3.34	0.06	0.41	0.27	0.18	0.30	0.31	0.09
MI	1	0.68	2.15	0.28	0.89	2.39	0.85	0.55	1.08	0.65
	2	1.70	3.43	0.08	0.82	1.15	0.59	0.93	0.98	0.52
	3	1.55	2.47	0.36	ND	ND	ND	ND	ND	ND
	4	ND	ND	ND	1.20	1.13	0.61	0.84	0.85	0.72
	5	3.14	3.49	0.38	1.30	0.64	0.41	1.35	0.91	0.43
MN	1	0.77	3.85	0.08	0.46	1.78	0.14	0.12	0.36	0.07
	2	2.27	5.58	0.17	0.34	0.65	0.14	0.18	0.33	0.03
	3	2.23	3.06	0.20	0.39	0.53	0.07	0.18	0.21	0.07
	4	2.59	4.97	0.24	0.33	0.34	0.12	0.23	0.28	0.05
	5	3.61	3.50	0.25	0.21	0.21	0.08	0.22	0.24	0.06
NE	1	2.90	5.00	ND	0.68	1.31	ND	0.95	1.70	ND
	2	0.88	1.39	ND	0.07	0.12	ND	0.25	0.42	ND
	3	4.00	3.89	ND	0.03	0.03	ND	0.03	0.03	ND
	4	4.62	4.53	ND	0.20	0.19	ND	0.18	0.18	ND
	5	2.55	3.82	ND	0.14	0.10	ND	0.06	0.07	ND
OR	1	ND	ND	ND	ND	ND	ND	0.03	0.03	0.03
	2	ND	ND	ND	ND	ND	ND	0.23	0.40	0.12
	4	1.77	1.06	0.27	ND	ND	ND	0.03	0.03	0.03
	5	3.36	2.75	1.05	0.36	0.34	0.26	0.55	0.66	0.41
UT	1	1.95	9.23	0.33	ND	ND	ND	0.21	0.63	0.11
	2	2.08	4.57	0.22	0.44	0.57	0.36	0.16	0.41	0.08
	3	3.33	5.68	0.30	0.56	0.49	0.50	0.14	0.23	0.06
	4	3.90	4.47	0.40	0.74	0.48	0.55	0.18	0.21	0.13
	5	6.58	6.13	0.57	1.47	2.79	0.65	0.17	0.32	0.11
WI	1	1.42	4.16	ND	0.33	0.74	0.21	0.27	0.47	0.17
	2	1.73	5.07	0.17	0.25	0.79	0.15	0.21	0.55	0.11
	3	1.43	5.62	0.29	0.36	0.79	0.27	0.20	0.45	0.14

*See Table 4 for description of soils.
†ND signifies not determined.

Table 6.   Concentrations of Zn in DTPA soil extracts and in leaf and grain of barley grown in 15 locations.

Location*	Year	DTPA Extract 20 mt/ha/y	100 mt/ha	NPK	Barley Leaf 20 mt/ha/y	100 mt/ha	NPK	Barley Grain 20 mt/ha/y	100 mt/ha	NPK
							--- mg/kg ---			
A2	1	17.9	59.5	5.1	46.4	58.1	30.6	51.5	59.2	46.7
	2	19.3	45.4	1.1	53.7	71.2	38.9	55.4	58.9	40.4
	3	44.2	52.9	4.4	35.5	36.7	16.4	55.7	66.4	46.7
	4	44.3	54.8	3.5	31.4	36.8	19.2	37.4	37.4	40.3
	5	46.6	55.1	3.3	ND†	ND	ND	ND	ND	ND
CA(G)	1	14.7	35.0	2.0	36.3	39.3	28.8	45.8	55.8	33.8
	2	18.3	23.6	1.8	22.0	29.5	20.8	34.0	40.5	26.3
	3	18.5	19.2	1.6	39.1	33.4	32.8	37.2	31.9	29.8
	4	21.4	15.8	1.6	32.4	24.5	26.6	33.3	39.4	28.9
	5	37.1	17.3	1.6	24.9	19.7	20.1	32.8	32.4	27.1
CA(D)	1	13.0	43.8	1.5	21.0	34.3	21.8	37.8	49.5	31.3
	2	34.0	55.9	2.3	25.0	27.8	22.5	34.3	37.0	37.3
	3	26.4	27.5	1.5	37.1	36.3	30.0	41.5	42.0	32.0
	4	26.5	22.4	1.7	25.8	23.2	20.1	30.6	33.7	25.8
	5	30.8	19.0	1.6	26.3	23.6	21.5	47.0	39.5	26.7
CO	1	ND	ND	ND	ND	ND	ND	57.8	58.8	51.8
	2	4.7	23.0	2.2	45.4	60.7	25.7	66.5	64.5	56.0
	3	6.7	28.0	2.8	ND	ND	ND	51.8	66.8	41.8
	4	14.5	38.9	2.4	ND	ND	ND	4.8	76.1	47.3
	5	18.9	40.0	2.6	ND	ND	ND	100.0	112.0	60.5
FL	1	11.3	74.5	1.7	23.9	33.4	13.9	13.5	18.0	10.7
	2	31.3	69.5	2.2	66.0	70.8	36.0	77.3	65.3	41.8
	3	48.2	76.5	2.5	153.1	93.3	38.5	55.3	63.8	43.0
	4	60.5	76.5	2.9	ND	ND	ND	67.1	73.8	30.7
	5	41.1	61.8	2.5	30.9	28.7	20.8	85.0	94.1	50.0
N	1	ND	119.3	3.6	ND	ND	ND	ND	82.1	57.5
	2	ND	33.1	1.3	ND	ND	ND	ND	60.2	37.1
	3	ND	31.0	2.0	ND	ND	ND	ND	61.6	35.8
	4	ND	41.4	3.3	ND	31.4	21.0	ND	43.5	27.0
	5	ND	68.5	2.3	ND	42.9	21.8	ND	44.0	31.8
MD(L)	1	ND	ND	ND	25.3	69.9	17.4	38.6	52.1	23.6
	2	9.3	53.6	2.4	21.4	58.7	17.0	32.2	46.7	27.6
	3	10.4	35.0	3.0	26.3	32.6	21.9	37.4	38.9	25.0
	4	23.7	39.5	3.4	54.0	112.4	17.2	55.6	66.1	34.4
	5	32.5	33.1	1.9	20.0	23.0	14.9	35.1	40.7	21.2
MD(H)	1	ND§	ND	ND	25.0	74.9	18.5	33.0	49.7	24.4
	2	8.4	54.5	1.6	19.6	32.8	11.3	27.9	38.2	22.1
	3	9.6	32.4	1.5	22.4	26.7	17.4	ND	30.6	20.5
	4	23.1	31.3	1.8	40.7	43.0	13.4	45.3	43.8	27.6
	5	64.5	30.0	1.3	20.9	23.4	14.7	33.4	35.1	16.5
MI	1	8.4	25.8	1.5	67.9	112.3	48.7	64.8	78.5	47.5
	2	21.2	44.3	1.8	48.3	63.2	27.9	110.0	85.3	69.5
	3	24.3	35.7	3.9	ND	ND	ND	ND	ND	ND
	4	ND	ND	ND	69.2	52.1	43.4	65.7	64.7	46.8
	5	52.8	56.4	5.4	79.5	57.5	44.0	118.3	155.0	121.5
MN	1	32.5	162.7	5.0	46.3	106.0	24.3	67.0	111.8	48.5
	2	34.0	85.5	2.0	38.8	52.0	28.5	67.5	80.3	41.3
	3	33.5	43.3	2.1	41.1	44.3	28.9	69.3	70.1	52.0
	4	35.8	67.4	2.8	45.0	57.7	40.7	49.4	46.4	36.0
	5	73.7	56.7	2.3	39.2	34.3	26.5	65.0	67.1	49.6
NE	1	31.5	102.0	ND	39.0	96.0	ND	94.0	118.0	ND
	2	18.1	54.8	ND	19.0	26.7	ND	98.1	111.0	ND
	3	45.5	43.8	ND	ND	ND	ND	64.3	63.5	ND
	4	59.4	62.4	ND	22.7	18.9	ND	75.5	69.5	ND
	5	54.8	52.6	ND	25.6	24.7	ND	49.7	47.7	ND
OH	1	23.9	69.3	9.4	30.0	40.5	26.0	53.0	70.3	47.3
	2	37.8	70.2	9.5	32.2	32.3	24.8	50.1	49.9	35.8
	3	41.9	49.2	8.3	31.0	28.2	24.7	36.0	36.0	28.1
	4	55.8	36.0	19.9	35.6	35.9	29.6	38.0	32.6	27.8
	5	43.6	21.9	10.1	19.5	15.6	18.6	24.8	23.6	19.9
OR	1	ND	ND	ND	ND	ND	ND	18.3	21.9	13.4
	2	ND	ND	ND	ND	ND	ND	56.0	70.0	52.5
	4	31.4	22.3	8.8	ND	ND	ND	53.3	57.8	43.8
	5	55.3	40.2	18.9	56.3	41.1	57.0	68.2	57.6	63.0
UT	1	29.0	129.8	2.5	ND	ND	ND	52.0	99.5	48.5
	2	29.9	58.5	2.8	26.0	32.3	21.5	65.3	77.5	47.5
	3	46.5	75.8	3.6	29.3	29.3	43.3	61.5	65.0	44.0
	4	50.4	57.5	5.4	37.3	35.8	22.0	65.5	58.8	71.0
	5	78.6	72.5	4.8	52.5	108.0	57.5	57.8	62.3	68.3
WI	1	29.6	74.6	3.5	61.2	80.4	54.4	54.4	78.4	36.5
	2	30.0	89.8	3.7	38.0	71.8	57.8	57.8	86.4	36.0
	3	23.6	95.1	3.9	28.5	42.1	53.5	53.5	68.8	40.3

*See Table I-4 for description of soils.
†ND signifies not determined.

Table 7. Concentrations of Cu in DTPA soil extracts and in leaf and grain of barley grown in 15 locations.

Location*	Year	DTPA Extract 20 mt/ha/y	100 mt/ha	NPK	Barley Leaf 20 mt/ha/y	100 mt/ha	NPK	Barley Grain 20 mt/ha/y	100 mt/ha	NPK
							- - mg/kg - -			
AZ	1	6.4	19.5	7.1	8.6	9.0	7.6	9.1	9.6	8.9
	2	8.3	15.5	2.4	10.4	12.1	9.9	10.8	11.8	9.5
	3	13.8	17.7	3.8	12.4	11.8	12.4	13.4	14.1	17.2
	4	13.8	15.7	2.8	10.9	11.8	11.9	12.7	13.2	13.5
	5	14.6	15.7	2.5	ND†	ND	ND	ND	ND	ND
CA(G)	1	4.7	20.5	1.0	8.3	8.4	6.4	4.4	4.2	2.6
	2	5.5	6.1	1.1	6.6	7.0	5.7	2.5	3.0	1.9
	3	5.8	5.3	1.0	7.8	6.1	6.2	3.9	3.6	3.6
	4	6.0	4.3	0.9	12.0	9.0	8.4	4.5	4.7	2.8
	5	9.7	4.3	0.8	10.1	8.8	8.4	9.6	10.1	11.3
CA(D)	1	4.9	26.8	1.4	6.7	9.1	6.1	3.6	4.0	3.2
	2	10.1	17.9	1.8	6.9	7.0	6.5	3.2	3.9	4.0
	3	8.9	10.3	1.4	10.4	8.8	7.7	3.7	3.4	3.2
	4	7.8	7.1	1.4	8.3	6.3	7.7	4.6	6.0	5.4
	5	9.2	5.9	1.2	11.4	10.9	12.2	18.4	17.0	15.8
CO	1	ND	ND	ND	ND	ND	ND	12.5	15.9	13.8
	2	3.2	9.1	2.3	9.5	11.1	9.6	44.4	39.7	44.8
	3	4.2	12.3	2.9	ND	ND	ND	6.4	5.3	5.3
	4	7.2	17.2	2.5	ND	ND	ND	6.7	6.2	5.6
	5	6.1	18.9	2.0	ND	ND	ND	14.2	13.1	8.2
FL	1	4.1	26.5	0.5	5.5	7.2	3.0	ND	ND	ND
	2	10.5	26.2	1.3	9.4	10.8	5.4	4.8	5.2	3.2
	3	13.9	19.8	0.6	11.8	9.3	4.7	4.0	4.1	1.8
	4	14.7	23.5	0.7	ND	ND	ND	ND	2.4	0.5
	5	15.7	22.8	0.5	6.9	5.7	4.7	6.5	6.4	2.9
IN	1	ND	30.4	4.1	ND	ND	ND	ND	5.7	4.6
	2	ND	3.8	2.0	ND	ND	ND	ND	7.4	6.1
	3	ND	13.7	3.2	ND	ND	ND	ND	5.9	4.7
	4	ND	18.0	3.7	ND	5.1	3.3	ND	5.0	3.4
	5	ND	24.7	3.5	ND	6.7	5.6	ND	7.1	6.8
MD(L)	1	ND	ND	ND	8.9	13.1	5.0	4.9	3.9	2.4
	2	3.7	1.8	1.1	5.9	9.7	5.4	4.4	5.7	3.6
	3	4.5	14.9	1.3	5.7	7.2	5.4	7.7	4.4	3.1
	4	11.8	17.6	1.8	11.5	12.5	6.3	6.1	2.9	4.5
	5	13.1	13.6	1.0	6.9	5.5	3.9	5.5	5.0	2.4
MD(H)	1	ND	ND	ND	8.9	14.2	6.7	5.2	5.1	3.6
	2	4.2	19.7	0.9	6.2	7.6	4.5	2.6	4.5	3.7
	3	4.7	13.4	0.9	5.5	6.6	4.9	ND	5.9	3.3
	4	11.7	17.5	1.2	12.3	12.8	6.4	5.4	7.0	4.6
	5	14.0	14.4	0.9	5.4	4.9	4.1	4.5	5.2	2.8
MI	1	2.8	9.6	0.6	27.9	27.4	31.3	22.3	14.0	26.0
	2	9.2	19.8	0.7	13.4	14.0	13.2	19.0	23.8	22.4
	3	7.5	11.5	1.5	ND	ND	ND	ND	ND	ND
	4	ND	ND	ND	37.0	25.0	27.3	18.0	16.1	21.1
	5	15.3	17.1	2.0	17.9	18.3	13.6	32.4	19.3	16.5
MN	1	8.8	43.3	2.0	10.4	11.1	16.8	0.5	0.5	3.8
	2	8.5	21.8	1.0	5.7	7.1	4.8	0.5	0.5	0.5
	3	8.1	11.4	0.9	6.9	7.4	6.7	2.5	2.0	3.4
	4	8.4	16.8	1.1	8.8	10.1	7.4	3.2	3.0	3.0
	5	15.8	11.3	1.3	7.4	9.8	7.2	6.0	5.9	5.9
NE	1	8.6	24.3	ND	5.6	8.3	ND	10.2	9.8	ND
	2	4.9	13.2	ND	5.8	7.2	ND	8.7	13.6	ND
	3	10.3	11.6	ND	6.8	9.0	ND	4.3	4.3	ND
	4	14.2	15.2	ND	4.8	5.0	ND	4.1	5.6	ND
	5	14.1	15.9	ND	4.7	4.6	ND	2.6	3.6	ND
OH	1	8.0	19.2	2.9	26.8	34.3	24.0	5.4	6.4	5.1
	2	13.1	27.9	2.5	19.4	16.1	22.1	7.3	7.2	5.7
	3	13.5	20.0	2.7	21.9	18.1	18.6	4.8	4.1	3.8
	4	14.2	16.0	3.5	23.9	27.3	25.8	3.6	3.3	3.0
	5	14.7	7.8	3.0	5.8	5.5	5.5	2.7	2.7	2.6
OR	1	ND	ND	ND	ND	ND	ND	4.5	5.0	2.3
	2	ND	ND	ND	ND	ND	ND	ND	ND	ND
	4	9.5	7.0	2.2	ND	ND	ND	4.2	5.1	2.4
	5	16.9	14.2	4.7	9.5	7.5	8.1	5.0	4.8	4.5
UT	1	13.2	54.0	1.4	ND	ND	ND	7.0	12.8	5.8
	2	12.6	24.2	1.3	9.7	9.5	8.3	7.6	7.6	5.7
	3	17.4	27.6	1.3	9.7	9.8	13.0	6.1	6.3	5.2
	4	19.3	22.6	2.0	12.6	11.4	8.8	6.0	5.7	5.2
	5	30.3	28.8	1.6	10.1	13.1	8.5	5.4	6.0	4.8
WI	1	9.7	24.8	1.3	12.4	12.7	9.6	7.1	8.6	5.5
	2	10.1	29.4	1.7	9.4	12.2	7.7	5.5	7.5	4.8
	3	6.9	28.7	1.9	10.5	11.0	9.2	5.9	6.8	4.9

*See Table 4 for description of soils.
†ND signifies not determined.

Table 8.   Concentrations of Ni in DTPA extracts and in leaf and grain of barley grown in 15 locations.

Location*	Year	DTPA Extract			Barley Leaf			Barley Grain		
		20 mt/ha/y	100 mt/ha	NPK	20 mt/ha/y	100 mt/ha	NPK	20 mt/ha/y	100 mt/ha	NPK
							mg/kg			
AZ	1	4.9	7.9	1.3	4.1	4.6	3.2	3.5	4.5	2.9
	2	1.5	3.3	0.1	2.2	2.3	1.3	6.5	6.5	5.1
	3	2.5	2.3	0.7	7.7	7.2	7.2	1.7	1.7	1.4
	4	2.3	2.2	0.7	10.2	4.5	5.2	1.7	1.7	1.4
	5	2.4	2.7	0.6	ND†	ND	ND	ND	ND	ND
CA(G)	1	1.8	7.6	0.6	ND	ND	ND	ND	ND	ND
	2	2.5	2.9	0.8	0.1	0.1	0.1	0.1	0.1	0.1
	3	2.2	1.9	0.4	2.0	2.0	2.0	2.0	2.0	2.0
	4	1.9	1.2	0.4	ND	ND	ND	ND	ND	ND
	5	4.1	1.7	0.7	1.0	1.0	1.0	1.0	1.0	1.0
CA(D)	1	1.6	9.3	0.6	0.1	0.1	0.1	0.1	0.1	0.1
	2	2.9	5.8	0.6	0.1	0.1	0.1	0.1	0.1	0.1
	3	2.4	2.5	0.4	2.0	2.0	2.0	2.0	2.0	2.0
	4	1.8	1.3	0.4	ND	ND	ND	ND	ND	ND
	5	3.0	1.8	0.8	1.0	1.0	1.0	1.0	1.0	1.0
CO	1	ND	ND	ND	ND	ND	ND	1.6	1.4	1.4
	2	0.8	2.2	0.5	1.3	1.4	0.7	1.6	2.0	2.1
	3	0.9	2.8	0.7	ND	ND	ND	0.9	0.8	0.8
	4	1.9	4.8	0.6	ND	ND	ND	0.7	1.1	0.6
	5	1.4	7.8	0.6	ND	ND	ND	0.7	3.7	1.0
FL	1	1.3	19.3	ND	ND	ND	ND	ND	ND	ND
	2	2.3	5.3	0.2	ND	ND	ND	ND	ND	ND
	3	2.8	5.0	0.2	ND	ND	ND	ND	ND	ND
	4	5.7	5.6	0.4	ND	ND	ND	ND	ND	ND
	5	4.2	4.5	0.2	0.1	0.1	0.1	0.1	0.1	0.1
IN	1	ND	43.3	6.8	ND	ND	ND	ND	2.7	1.4
	2	ND	4.0	7.8	ND	ND	ND	ND	1.0	0.4
	3	ND	6.9	1.9	ND	ND	ND	ND	0.6	0.3
	4	ND	2.0	1.7	ND	1.1	1.3	ND	2.9	1.4
	5	ND	2.2	1.8	ND	0.7	0.5	ND	0.8	0.3
MD(L)	1	ND	ND	ND	0.3	0.5	0.5	0.5	1.0	0.2
	2	1.6	7.2	1.4	2.6	2.8	2.6	0.6	0.7	0.6
	3	2.3	4.2	1.0	1.0	1.0	1.5	0.4	0.5	0.4
	4	6.3	7.8	3.3	1.8	1.9	1.9	0.6	0.7	0.4
	5	4.5	5.2	0.9	1.2	1.5	0.9	1.0	1.3	0.6
MD(H)	1	ND	ND	ND	0.3	0.4	0.5	6.0	0.7	0.5
	2	1.7	5.7	0.6	2.5	2.6	2.6	0.4	0.6	0.4
	3	1.3	2.9	0.7	1.2	1.2	1.3	-1.0	0.3	0.6
	4	5.0	5.0	3.0	2.0	2.2	2.0	0.5	0.6	0.3
	5	3.8	3.8	1.1	1.0	1.1	1.2	1.3	1.1	0.7
MI	1	0.9	3.8	0.2	1.2	2.3	1.2	11.2	5.3	13.0
	2	2.7	5.7	0.5	1.3	1.5	1.4	1.2	1.1	0.5
	3	2.3	3.4	3.3	ND	ND	ND	ND	ND	
	4	ND	ND	ND	1.1	1.0	1.1	0.3	0.4	0.2
	5	3.8	4.4	0.5	0.9	1.0	0.7	1.3	0.9	0.5
MN	1	7.5	24.3	4.1	1.1	1.3	1.3	0.4	2.7	0.3
	2	6.5	12.3	2.3	0.6	0.9	0.5	1.4	3.2	1.2
	3	4.6	5.4	2.0	0.7	0.8	1.0	0.9	1.1	0.9
	4	6.2	11.2	2.4	1.1	2.0	1.1	0.9	1.5	0.8
	5	9.7	7.8	2.8	0.8	0.7	0.4	1.1	0.8	0.8
NE	1	3.9	6.5	ND	2.8	5.0	ND	1.6	1.6	ND
	2	3.5	8.1	ND	1.1	1.1	ND	1.2	4.8	ND
	3	11.6	7.8	ND	1.2	5.0	ND	0.3	0.5	ND
	4	5.7	5.8	ND	1.2	1.1	ND	0.9	1.0	ND
	5	11.1	9.6	ND	10.1	6.3	ND	0.8	1.2	ND
OH	1	1.9	6.7	0.5	1.5	1.7	0.9	0.5	1.2	0.5
	2	3.5	8.7	1.0	1.0	2.3	1.4	0.6	0.7	0.5
	3	3.2	4.3	0.8	2.0	1.7	1.5	0.5	0.3	0.2
	4	3.7	3.5	0.7	0.6	0.5	0.5	0.3	0.2	0.1
	5	3.7	1.8	0.9	0.9	1.0	0.6	0.4	0.5	0.7
OR	1	ND	ND	ND	ND	ND	ND	3.8	5.4	2.5
	2	ND	ND	ND	ND	ND	ND	2.0	2.0	2.0
	4	4.4	2.8	1.3	ND	ND	ND	2.5	4.7	2.5
	5	6.4	4.2	2.2	0.1	1.3	0.1	0.1	1.3	0.1
UT	1	3.4	12.7	0.7	ND	ND	ND	0.3	0.9	0.2
	2	3.0	5.6	0.6	2.0	2.3	1.9	0.2	0.3	0.1
	3	4.5	6.4	0.7	3.8	3.6	4.6	0.1	0.1	0.1
	4	4.3	4.5	0.6	2.3	1.6	2.0	0.3	0.4	0.1
	5	3.9	5.3	0.5	1.8	2.3	1.6	0.2	0.2	0.1
WI	1	8.6	14.6	5.5	3.1	3.5	3.4	2.7	4.2	1.9
	2	8.4	14.9	5.0	2.2	2.8	2.0	1.8	2.8	1.4
	3	7.9	14.8	5.4	0.8	1.0	0.7	0.5	1.3	0.4

*See Table 4 for description of soils.
†ND signifies not determined.

SOIL PROPERTIES INFLUENCING THE ACCUMULATION
OF TRACE ELEMENTS BY PLANTS

The accumulation of trace elements by plants is a reflection
of the influence of soil physical properties on plant growth,
soil hydraulic properties and chemical properties such as pH,
CEC, and clay mineral sorption reactions.

Physical Properties

Soil particle size distribution (i.e., texture), structure,
and depth are important in determining soil hydraulic properties
such as porosity, permeability and drainage rates; these properties in turn influence soil moisture content and aeration/
respiration which impact the type and rates of both soil microbial activity and chemical reactions, as well as plant root
development and growth rates.

Although soil texture, hard pans, and other physical features
can be observed in the field or identified from soil maps and
likely influence soil chemical reactions, clear identification
of these effects in relation to plant uptake of metals has been
difficult. Soil texture, however, has been recommended as a
quantifiable soil property to limit metal loadings to soils in
the Northeast (Baker et al., 1985), although experimental data
are not available to support this concept. Presumably as the
texture of the soil becomes finer (i.e., greater clay content),
the limiting application rates may also increase.

Soil pH

The impact of pH on metal accumulation by plants has been
extensively reviewed (Logan and Chaney, 1983) and little additional pertinent data has been reported which would refute their
conclusions. Basically, metal availability (except for Mo and
Se) tends to decrease with liming. Solubility of solid phase
minerals including metal carbonates, phosphates, and sulfides
is enhanced at low soil pH (Lindsay, 1979); however, the importance of this phenomenon in sewage sludge-treated soils has not
been adequately defined by solid phase equilibria studies.

Soil pH is one of the easier characteristics to measure but
care must be exercised in interpreting results. The pH measured
in a soil-neutral salt suspension will be lower than the pH
measured in a soil-water system, although some exceptions exist.
When pH was measured at 19 sites over a growing season, a maximum
variation of 1.6 pH units was reported for measurements in a 1:2
soil:water suspension or in 0.01 $\underline{N}$ $CaCl_2$ (Collins et al., 1970).
Bates et al. (1982) found similar variations in Ontario, Canada,
soils, whether pH was measured in 0.01 $\underline{N}$ $CaCl_2$ or in a soil-water
saturated paste. These authors found greater variation in the
pH for soils cropped to corn which received large applications
of N fertilizer than for soils cropped to alfalfa.

The water content of the soil and its electrolyte content affects soil pH readings significantly (Thomas and Hargrove, 1984). Bates et al. (1982) measured the pH in 245 soils and obtained the following pH values: pH 5.77 ± 0.91 for a soil: water saturated paste; pH 5.26 ± 0.91 for a 1:2 soil:solution of 0.01 N $CaCl_2$; and pH 5.00 ± 0.95 for a 1:2 soil:solution of 1 N KCl. The differences between methods varied with pH, such that: $pH_{H_2O} = 0.82 \pm 0.94$ pH_{CaCl_2}, $R^2 = 0.904$. This variability emphasizes the need for use of a standardized method to measure soil pH.

Soil organic matter and the resultant impact on pH buffering can influence the effect of liming on trace element uptake. Liming acid soils to pH 6.5 as measured in water paste (1:1 soil-water ratio), often is costly and can require considerable amounts of lime. Some trace element deficiencies (e.g., Fe, and Mn) may occur as pH approaches neutrality. In addition, the increase in soil pH may not reduce markedly the uptake of trace elements from sludge treated soils, especially for crops not accumulating metals. Pepper et al. (1983) attributed the ineffectiveness of liming on reducing Cd uptake by corn to a drop in soil pH during the growing season. Others (Hemphill et al., 1982; Giordano and Mays, 1981) have observed similar effects on metal content of corn grain. In general, lime applications reduce uptake of Zn and Ni more than Cd (Singh and Narwal, 1984).

Metal uptake by plant species may vary in response to liming. Giordano (CAST, 1980) observed that liming reduced Zn concentrations in soybean seed to a greater extent than in corn grain or cotton seed. Whereas the Ni content of soybean and cotton seed was depressed by lime, the content of Cu in all crops and plant parts was relatively unaffected.

Soil pH influences uptake of most metals at least to some extent, but the current recommendation of pH 6.5 should be reconsidered for food-chain agricultural soils since some reports indicate adequate control of metal uptake at pH 6. For example, Hajjar (1985) conducted a greenhouse study using soils treated with sludge at rates from 0 to 27 mt/ha. Replicate soils were adjusted to pH <5.6, pH 5.7-6.0, and pH >6.4 with either H_2SO_4 or $Ca(OH)_2$. For both tobacco and peanut, plant tissue concentrations of Ni, Zn, and Cd decreased with increasing soil pH; however, in general, metal uptake by peanuts was similar at pH >6.4 and pH 5.7-6.0, suggesting that attainment of pH 6.5 is not essential for minimizing plant availability of Ni, Zn and Cd. A soil pH ≈6.0 may be equally acceptable to regulate metal uptake. This conclusion is supported by field data with barley (Vlamis et al., 1985). More specific pH values where accumulator crops such as tobacco are grown on highly buffered acid soils may be required.

It has not been demonstrated that pH control to prevent transfer of metals into the food chain is necessary on forested sites. If a forest site is shifted to agricultural use or residential development, the soil pH should be adjusted by limestone addition at that time to meet existing standards.

Iron

Iron deficiency chlorosis on calcareous soils is a unique soil fertility problem worldwide. Application of sewage sludges can correct Fe chlorosis problems (McCaslin and O'Connor, 1985; McCaslin et al., 1986). In New Mexico, where Fe chlorosis affects large acreages of farmland, sludges applied at 34 to 90 mt/ha increased the levels of plant available Fe, Zn, and P in a severely Fe-deficient calcareous soil. Sorghum grain yields from sludge-treated soils were significantly higher than those receiving dairy manure or chemical fertilizers. Uptake of Zn and Cd by barley was minimal after sludge application to calcareous soils in the regional W-124 study (Tables 5 to 8).

Molybdenum

Soon and Bates (1985) present data which show that the application of a lime-treated sewage sludge supplying 0.21 kg Mo per hectare per year raised the Mo concentration in both bromegrass and corn stover above the control by significant margins (0.29 vs. 1.16 mg/kg for bromegrass, and 0.20 vs. 0.47 mg/kg for corn stover). A non-significant increase occurred in the Mo content of corn grain even though the lime-treated sludge increased soil pH from 7.4 to 8.1. In the same experiment, application of Al- and Fe-treated sludges raised the Mo concentration in bromegrass from 0.29 to 0.69 and 0.46 mg/kg, respectively. The amount of Mo added with the Al- and Fe-treated sludges averaged 2.18 and 1.66 kg per hectare per year, respectively.

Pierzynski and Jacobs (1986b) applied a sludge containing 1500 mg Mo/kg at rates of 42 and 94 mt/ha (equivalent to 63 and 141 kg Mo/ha). During the three year study the Mo content of corn seedlings (25-31 cm height) ranged from 47 to 724 mg/kg for those grown with the higher sludge application as compared to a range of 1.9-6.0 mg/kg in those from control plots. Similar increases were observed for soybean seedling tissue (18-23 cm height) and diagnostic leaf tissue of both crops. At the end of the study, soil pH had increased from 4.6 in the control to 6.9 in the 94 mt/ha sludge treatment. This change in soil pH may explain the increase of Mo concentration in plant tissue. The Mo content of corn grain also increased with time (0.2 to 0.6 mg Mo/kg for control and 2.0 to 6.9 mg Mo/kg for the high sludge rate) but the effect on the Mo concentration in soybean seeds was greater than on corn grain (8.9 to 19.9 mg Mo/kg for control and 122 to 242 mg Mo/kg for the high sludge rate). Elevated Mo levels did not affect growth of either crop.

A greenhouse study has shown that uptake of Mo by ryegrass and white clover was enhanced more by addition of up to 0.41 kg Mo/ha from sludge than from Na molybdate (Williams and Gogna, 1983). However, Mo additions from sludge did not always enhance Mo uptake by corn, soybeans, and alfalfa compared to Na molybdate, when rates of Mo from 60 to 400 kg Mo/ha were used (Pierzynski and Jacobs, 1986a). Results from another study indicated that sludge applications, especially to high pH soils, tended to reduce the Cu/Mo ratio of the affected vegetation (Soon and Bates, 1985). Under these conditions the likelihood of Mo induced Cu deficiency in grazing ruminant animals consuming the forage is enhanced.

Selenium

Sludge application to agricultural soils did not increase Se uptake by crops (Dowdy et al., 1984; Logan et al., 1987). In these studies, Se inputs ranging from 0.024 to 66 kg/ha did not result in significant accumulation of Se in plant tissue. Unless additional data become available to indicate otherwise, Se should not be a limiting factor in land application of municipal sludges.

Cation Exchange Capacity

Cation exchange capacity has been used as the primary soil property to govern metal loadings for the past 10-15 years. The basic concept originated in England and was adopted to prevent metal toxicities to crops; however, its use was mainly intended for soils where organic matter contributes a significant fraction to the CEC.

Method of Analysis

One problem when using CEC to regulate sewage sludge addition to soil is that no single method of determining CEC is universally accepted. The two most widely used methods are: 1) summation of exchangeable cations, and 2) saturation with either a buffered or unbuffered index cation. The above mentioned methods can give vastly different CEC values for the same soil. Hence, the recommended total metal loading rate and subsequent metal uptake by plants can vary depending upon the method used to determine CEC.

Criteria developed under 40 CFR part 257 of the Resource Conservation and Recovery Act (RCRA) state that the method to be used for CEC analysis should depend upon the type of soil (EPA, 1979). For distinctly acid soils the summation method should be used and for neutral, calcareous, or saline soils the sodium acetate method should be used (see Rhoades, 1982). If CEC is to be used as an index for metal loadings, then the method of analysis must be standardized.

Correlation of CEC and Plant Uptake of Metals

Research on the relationship between CEC and plant uptake of metals has been minimal and results have been conflicting (CAST, 1980; Logan and Chaney, 1983). Hinesly et al. (1982) conducted a study to determine the effect of CEC on Cd uptake by corn.

Soil samples of the B_1 horizon of the Ava series and the Ap horizon of the Maumee series were separately mixed with samples of
the Plainfield series to obtain soil mixtures having a CEC from
5.3 to 15.9 cmol(+)/kg(meq/100 g). Additions of $CdCl_2$ or 100
mt/ha of dried, digested sewage sludge were used to provide a
soil-Cd concentration of 10 mg/kg. Corn was grown in pots of
each mixture and harvested at 3- and 7-week intervals for tissue
analyses. The soil CEC inversely affected the uptake of Cd by
corn when Cd was supplied as a soluble salt, but not when it was
supplied as a constituent of municipal sewage sludge. This
conclusion was confirmed in greenhouse studies conducted by
Korcak and Fanning (1985).

In terms of phytotoxicity, research data available indicates
that the maximum metal loadings allowed in the CEC-metal limit
approach are conservative. Furthermore, a large degree of safety
is provided by the CEC-metal limit approach. For example, no
phytotoxicities have occurred in studies where the total metal
loading equals or greatly exceeds those recommended in the CEC
table at pH 6.5 (Chang et al., 1983; Ellis et al., 1981; Hinesly
et al., 1984; Vlamis et al., 1985). These observations indicate
that the present practice of using CEC as a basis for establishing metal-loading limits should be abandoned.

CONCLUSIONS

The following conclusions are supported by previous literature or by new information reported in this total report.

1. Conclusions on the impact of sewage sludge on trace element
    uptake by plants should be based on field studies rather
    than greenhouse or pot studies. Plant tissue concentrations
    obtained in greenhouse or pot studies may not be represen-
    tative of those obtained in the field unless root growth is
    not restricted and accumulation of soluble salts is avoided.

2. Concentrations of trace elements in crops grown in soils
    treated with metal salts exceed those of crops grown on
    sewage sludge-amended soils and should not be used to predict
    dietary intake.

3. Soil physical properties are related to trace element uptake
    by plants but current data does not allow quantification of
    these relationships.

4. Due to natural variations in soil and crop characteristics,
    trace element content of crops grown on untreated soils
    differ.

5. The soil pH value used in conjunction with application of
    sewage sludge should be measured on the sewage sludge-soil
    mixture or on the untreated soil using a 1:1 soil water
    ratio, realizing that 0.01 M $CaCl_2$ is a preferred matrix to
    compensate for soluble salt levels in sewage sludge or soils.

6. The impact of the pH reduction on increased metal uptake is more marked with high metal sludges and crops responsive to metal additions.

7. The relationship of either CEC or texture to metal uptake in sewage sludge-amended soils has not been conclusively demonstrated under field conditions. The current guidelines that are based on the use of CEC to limit metal additions to soils are not supported by current long-term field experimentation.

8. Trace element deficiencies rather than toxicities are a major concern in soils containing free $CaCO_3$. Sewage sludge additions have been used to correct Fe deficiency in calcareous soils. Based on plant uptake, Mo is the principal metal of concern in calcareous soils treated with sewage sludge.

REFERENCES

Baker, D. E. and M. C. Amacher. 1981. The development and interpretation of a diagnostic soil testing program. Penn. Agr. Exp. Sta. Bull. 826.

Baker, D. E., D. R. Bouldin, H. A. Elliott, and J. R. Miller. 1985. Criteria and recommendations for land application of sludges in the Northeast. Penn. Agri. Exp. Sta. Bull. 851.

Bates, T. E., J. Hassink, M. McAlpine, L. Rowsell and E. F. Gagnon. 1982. Progress Report, Department of Land Resources Science, University of Guelph, Ontario, Canada.

Bingham, F. T. 1980. Nutritional imbalances and constraints to plant growth on salt-affected soils. Agron. Abstr. American Society of Agronomy. Madison, WI. p. 49.

Council for Agricultural Science and Technology (CAST). 1980. Effects of sewage sludge on the cadmium and zinc content of crops. Council for Agric. Sci. Technol., Ames, IA.

Chang, A. C., A. L. Page, J. E. Warneke, M. R. Resketo, and T. E. Jones. 1983. Accumulation of cadmium and zinc in barley grown on sludge-treated soils: A long-term field study. J. Environ. Qual. 12:391-397.

Collins, J. B., E. P. Whiteside and C. E. Cross. 1970. Seasonal variability of pH and lime requirements in several southern Michigan soils when measured in different ways. Soil Sci. Soc. Am. Proc. 34:56-61.

Davis, R. D. 1981. Copper uptake from soil treated with sewage sludge and its implication for plant and animal health. pp. 223-241. IN P.L. 'Hermite and J. Dehandtschutter (ed.), Copper in animal wastes and sewage sludge. Reidel Publ. Co., Dordrecht, Holland.

deVries, M. P. C. and K. G. Tiller. 1978. Sewage sludge as a soil amendment with special reference to Cd, Cu, Mn, Ni, Pb, and Zn—A comparison of results from experiments conducted inside and outside a greenhouse. Environ. Pollut. 16:231-240.

Dijkshoorn, W., J. E. M. Lampe, and L. W. van Breckhoven. 1981. Influence of soil pH on heavy metals in ryegrass from sludge-amended soil. Plant Soil 61:277-284.

Dowdy, R. H., R. D. Goodrich, W. E. Larson, B. J. Bray and D. E. Pamp. 1984. Effects of sewage sludge on corn silage and animal products. EPA 600-2-84-075. Municipal Environmental Research Laboratory, Office of Research and Development, U.S. Environmental Protection Agency, Cincinnati, OH.

Ellis, B. G., A. E. Erickson, L. W. Jacobs, J. E. Hook and B. D. Knezek. 1981. Cropping systems for treatment and utilization of municipal wastewater and sludge. EPA 600/2-81-065. R. S. Kerr Environmental Res. Lab., Office of Res. and Dev., U.S. Environmental Protection Agency, Ada, OK.

Environmental Protection Agency. 1979. Criteria for classification of solid waste disposal facilities and practices. Final, interim final, and proposed regulations. (40 CFR 257). Fed. Reg. 44(179): 53460-53468.

Giordano, P. M. and D. Mays. 1981. Plant nutrients from municipal sewage sludge. Ind. Eng. Chem. Prod. Res. Dev. 20:212-216.

Hajjar, L. M. 1985. Effect of soil pH and the residual effect of sludge loading rate on the heavy metal content of tobacco and peanut and the effect of $CdSO_4$ on the Cd content of various cultivars of grasses. M.S. thesis, North Carolina State Univ., Raleigh.

Hemphill, D. D., Jr., T. L. Jackson, L. W. Martin, G. L. Kiemnec, D. Hanson and V. V. Volk. 1982. Sweet corn response to application of three sewage sludges. J. Environ. Qual. 11:191-196.

Hinesly, T. D., K. E. Redborg, E. L. Ziegler, and J. D. Alexander. 1982. Effect of soil cation exchange capacity on the uptake of cadmium by corn. Soil Sci. Soc. Am. 46(3):490-497.

Hinesly, T. D., K. E. Redborg, R. I. Pietz and E. L. Ziegler. 1984. Cadmium and zinc uptake by corn (Zea mays L.) with repeated application of sewage sludge. J. Agric. Food Chem. 32:155-163.

Holmgren, G. G. S., M. W. Meyer, R. B. Daniels, R. L. Chaney and J. Kubota. 1987. Cadmium, lead, zinc, copper, and nickel in agricultural soils of the United States. J. Environ. Qual. (in press).

Korcak, R. F. and D. S. Fanning. 1985. Availability of applied heavy metals as a function of type of soil material and metal source. Soil Sci. 140:23-34.

Lindsay, W. L. 1979. Chemical equilibria in soils. John Wiley and Sons, New York.

Lindsay, W. L. and W. A. Norvell. 1978. Development of a DTPA soil test for zinc, iron, manganese and copper. Soil Sci. Soc. Am. J. 42:421-428.

Logan, T. J. and R. H. Miller. 1983. Background levels of heavy metals in Ohio farm soils. Ohio Agr. Res. and Dev. Ctr. Research Circular 275.

Logan, T. J. and R. L. Chaney. 1983. Utilization of municipal wastewater and sludge on land—Metals. p. 235-326. IN A. L. Page et al. (ed.), Proc. of the Workshop on utilization of municipal wastewater and sludge on land. University of California, Riverside, CA.

Logan, T. J., A. C. Chang, and A. L. Page. 1987. Accumulation of selenium in crops grown on sludge-treated soil. J. Environ. Qual. (in press).

Lund, L. J., E. E. Betty, A. L. Page and R. A. Elliott. 1981. Occurrence of naturally high cadmium levels in soils and its accumulation by vegetation. J. Environ. Qual. 10:551-556.

McCaslin, B. D., J.-G. Davis, L. C. Chacek, and L. A. Schluter. 1987. Sorghum yield and soil analysis from sludge amended calcareous iron deficient soil. Agron. J. 79:204-209.

McCaslin, B. D. and G. A. O'Connor. 1985. Potential fertilizer value of gamma irradiated sewage sludge on calcareous soils. New Mexico Agr. Exp. Sta. Bull. 692.

Page, A. L. 1974. Fate and effects of trace elements in sewage sludge when applied to agricultural lands. A literature review-study. U.S. EPA Report NO. EPA-670/2-74-005. 108 pp.

Pepper, I. L., D. F. Bezdicek, A. S. Baker and J. M. Sims. 1983. Silage corn uptake of sludge-applied Zn and Cd as affected by soil pH. J. Environ. Qual. 12:270-275.

Pierce, F. J., R. H. Dowdy and D. F. Grigal. 1982. Concentrations of six trace metals in some major Minnesota soil series. J. Environ. Qual. 11:416-422.

Pierzynski, G. M. and L. W. Jacobs. 1986a. Extractability and plant availability of molybdenum from inorganic and sewage sludge sources. J. Environ. Qual. 15(4):323-326.

Pierzynski, G. M. and L. W. Jacobs. 1986b. Molybdenum accumulation by corn and soybeans from a Mo-rich sewage sludge. J. Environ. Qual. 15(4):394-398.

Rhoades, J. D. 1982. Cation exchange capacity. In: A. L. Page, et al. (ed.) Methods of soil analysis, Part 2. 2nd. Ed. Agronomy 9:149-157.

Römheld, V., and H. Marschner. 1986. Evidence for a specific uptake system for iron phytosiderophores in roots of grasses. Plant Physiol. 80:175-180.

Shacklette, H. T., and J. E. Boerngen. 1984. Element concentrations in soils and other surficial materials of the conterminous United States. U.S. Geological Survey Professional Paper 1270. U.S. Government Printing Office, Washington, D.C.

Singh, B. R., and R. P. Narwal. 1984. Plant availability of heavy metals in a sludge-treated soils: II. Metal extractability compared with plant metal uptake. J. Environ. Qual. 13:344-349.

Soon, Y. K. 1981. Solubility and sorption of cadmium in soils amended with sewage sludge. J. Soil Sci. 32:85-95.

Soon, Y. K. and T. E. Bates. 1985. Molybdenum, cobalt and boron uptake from sewage sludge-amended soils. Can. J. Soil Sci. 65:507-517.

Thomas, G. W., and W. L. Hargrove. 1984. The chemistry of soil acidity. p. 40. In F. Adams (ed.) Soil acidity and liming. 2nd ed. Agronomy 12:3-56.

Vlamis, J., D. E. Williams, J. E. Corey, A. L. Page, and T. J. Ganje. 1985. Zinc and cadmium uptake by barley in field plots fertilized seven years with urban and suburban sludge. Soil Sci. 139:81-87.

Williams, J. H., and J. C. Gogna. 1983. Molybdenum uptake: Residual effect of sewage sludge application. p. 483-486. In Proc. 4th Int. Conf. on Heavy Metals in the Environment, Vol. 1, Heidelberg, 1983. CEP Consultants, Ltd., Edinburgh, UK.

Wolnik, K. A., F. C. Fricke, S. G. Capar, G. C. Braude, M. W. Meyer, R. D. Satzger and E. Bonnin. 1983a. Elements in major raw agricultural crops in the United States. 1. Cadmium and lead in lettuce, peanuts, potatoes, soybeans, sweet corn, and wheat. J. Agric. Food Chem. 31:1240-1244.

Wolnik, K. A., F. L. Fricke, S. G. Capar, M. W. Meyer, R. D. Satzger and R. W. Kuennen. 1983b. Elements in major raw agricultural crops in the United States. 2. Other elements in lettuce, peanuts, potatoes, soybeans, sweet corn and wheat. J. Agric. Food Chem. 31:1244-1249.

Wolnik, K. A., F. L. Fricke, S. G. Capar, M. W. Meyer, R. D. Satzger, E. Bonnin and C. M. Gaston. 1985. Elements in major raw agricultural crops in the United States. 3. Cadmium, lead, and eleven other elements in carrots, field corn, onions, rice, spinach, and tomatoes. J. Agric. Food Chem. 33:807-811.

CHAPTER 3

EFFECTS OF SLUDGE PROPERTIES ON
ACCUMULATION OF TRACE ELEMENTS BY CROPS

Richard B. Corey
University of Wisconsin, Madison, Wisconsin

Larry D. King
North Carolina State University, Raleigh, North Carolina

Cecil Lue-Hing
Metropolitan Sanitary District of Greater Chicago, Chicago, Illinois

Delvin S. Fanning
University of Maryland, College Park, Maryland

Jimmy J. Street
University of Florida, Gainesville, Florida

John M. Walker
U.S. Environmental Protection Agency, Washington, D.C.

INTRODUCTION

Loading limits for trace elements from municipal sewage
sludges applied to land should be based on sludge and soil
characteristics that affect plant availability of those elements.
Current evidence indicates that the rate at which a plant root
absorbs a trace element such as Cd, Zn, or Cu depends on the
activity of the free-ion form of that metal in solution at the
root surface. The activity at the root surface, in turn, depends
on equilibrium reactions between solution and solid phases and
the rate of transport to the root. Therefore, if we can predict
the trace element uptake that a specific application of a given
sludge on a particular soil will produce, we should be able to
establish long-term loading limits that will assure that additions of trace elements to the food chain are within tolerable
limits and phytotoxicities of other trace elements are not a
problem.

25

Sludges by themselves support certain trace element activities when equilibrated with the soil solution. Adsorbing sites on the soil immobilize some of the dissolved trace element ions, causing more ions to be released from the sludge. If the trace element adsorption capacity of the applied sludge is small compared to the adsorption capacity of the soil, the soil properties will be very important in determining the equilibrium solution activity. However, if the trace element adsorption capacity of the sludge is high compared to that of the soil (usually associated with high sludge rates), the soil adsorption sites that can be filled at the activity supported by the sludge will result in only a small decrease in solution activity, and the sludge properties will dominate. Under these conditions the soil's effect on the pH of the mixture may still be significant, and the pH will affect trace metal concentrations in equilibrium with the sludge.

The fact that trace element activity, and thus plant uptake of trace elements from sludged soils, tends to approach a maximum as the sludge rate increases suggests that this behavior could be used to differentiate sludges that would support potentially harmful concentrations in plant tissue from sludges that would not, regardless of the application rate. Therefore, trace element activity could be used to differentiate sludges that do and do not require regulation in terms of trace element loading limits. (Fig. 1). In this figure, a constant pH is assumed, and

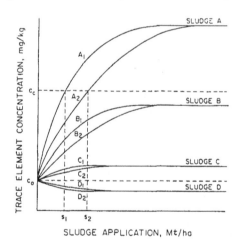

SLUDGE APPLICATION, Mt/ha

Figure 1. Basis for differentiating sludges that do not require loading limits to prevent harmful trace element accumulations in plants (sludges B, C, and D) from one that does (sludge A). The two curves for each sludge represent that sludge applied to a soil of relatively low adsorption capacity (subscript 1) and a soil of higher adsorption capacity (subscript 2). C_0, the concentration of a given element in plant grown on the unamended soil, is shown as being the same for both soils. C_c is the critical concentration in the plant, and S_1 and S_2 are loading limits for sludge A applied to soils (1) and (2), respectively.

the dashed horizontal line (C_c) represents the maximum concentration of trace element that would be allowed in a test species based on the maximum allowable dietary intake calculations or on phytotoxicity. In the case of Cd, this line would represent the maximum allowable concentration in a test plant (such as lettuce), based on the assumption that: (1) uptake by different plant species grown on soils amended with sludge will show proportionate differences in Cd uptake, and (2) the total intake from a "market basket" mix of species grown on a given sludge treatment can be estimated from the concentration determined in the test species. The letters A, B, C, and D represent Cd uptake curves derived from sludges that support different maximum Cd concentrations in plants, and the subscripts 1 and 2 represent uptake curves for the designated sludge applied to soils with low and high Cd-adsorption properties, respectively. If the maximum plant-Cd concentration that sludge-soil mixture can support exceeds the critical concentration in the plant, C_c, the permissible sludge loading would depend on soil properties (Curves A_1 and A_2). Sludges B, C, and D, that maintain plant Cd below the critical level regardless of sludge loading at or above the pH used in the test, would not require loading limits regardless of the soil properties. In fact, sludge D represents a sludge that supports a lower plant Cd concentration than does the nonamended soil. This is a rare occurrence, but it has been observed.

Estimates of plant availability of the trace elements in sludges could be obtained from field or greenhouse studies with a specific variety of lettuce (or other crop that tends to accumulate the elements of interest) grown on sludge-amended soils that had been allowed to equilibrate under aerobic conditions. Later, reliable methods for assessing trace element supplying properties of sludges and adsorption characteristics of soils may be developed for use in place of the bioassay. Interpretation of such tests will require research relating test results to plant uptake. Use of chelating resins for desorbing trace elements from sludges and for establishing known trace element activities for soil adsorption curves appears promising.

FORMS AND AMOUNTS OF TRACE ELEMENTS IN MUNICIPAL SEWAGE SLUDGES

All domestic sewage sludges contain varying amounts of Cd, Cr, Cu, Pb, Ni, and Zn. Data presented by Sommers (1977) showed wide variations in metal concentrations and a fairly large number of sludges containing very high concentrations. With the implementation of the federal industrial waste pretreatment program as a control on the discharge of these trace metals into publicly-owned treatment facilities, the metal loads to municipal wastewater treatment facilities and subsequently the levels in municipal sludges can be expected to decrease with time. The federal program, which is likely to generate vast amounts of performance data in the future, is not mature enough to produce such data at this time. However, the impact of this program can be illustrated by examining trends in sludge metal composition data from Chicago, Baltimore, and Philadelphia where local pretreatment programs have been implemented.

Trends in sludge trace-metal concentrations

The Metropolitan Sanitary District of Greater Chicago (MSDGC) adopted a program in 1969 that was designed with objectives similar to those of the federal program which prevents pass-through of pollutants and produces higher quality effluents. The program specifies concentration limits for 13 contaminants and nine limiting conditions for the discharge of industrial wastes to the MSDGC sewerage system. The specific limits required by this program are as follows (in mg/L):

boron, 1.0; cadmium, 2.0; chromium (total), 25.0; chromium (hexavalent), 10.0; copper, 3.0; cyanide (total), 10.0; cyanide (readily releasable at 150°F and pH 4.5), 2.0; iron, 50; lead, 0.5; mercury, 0.005; nickel, 10; zinc, 15.0; fats, oils and greases, 100 (changed to 250 in 1983); and pH 4.5 to 10 units.

The levels of sewage-borne metals entering the MSDGC sewage treatment facilities (Tables 9 and 10) have decreased

Table 9. Metal loadings and cumulative percent reductions to Chicago area treatment facilities, 1971 through 1977.[*]

	Cd	Cr	Cu	Pb	Ni	Zn
	— — — — — — — — — — kg/day — — — — — — — — — —					
1971	398	5,197	2,166	2,049	2,443	6,972
1972	343	3,321	1,996	1,793	1,377	4,641
1973	301	2,463	961	1,063	957	4,260
1974	213	1,894	652	735	643	3,403
1975	113	1,522	538	497	386	2,537
1976	132	1,527	685	368	416	2,400
1977	168	1,422	588	536	436	2,587
Cumulative % reduction	57.7	72.6	72.9	73.8	82.2	62.9

[*]From Whitebloom et al. (1978).

Table 10. Metal loadings and cumulative percent reductions to Chicago area treatment facilities, 1971 through 1984.[*]

	Cd	Cr	Cu	Pb	Ni	Zn
	— — — — — — — — — — kg/day — — — — — — — — — —					
1971	398	5,197	2,166	2,049	2,443	6,972
1984	121	1,185	949	396	702	2,322
Cumulative % reduction	69	77	56	81	71	67

[*]Lue-Hing (1985, personal communication). Data in Tables 9 and 10 are from two different POTWs within the MSDGC system.

substantially since 1971 (Whitebloom et al., 1978; Lue-Hing, 1985, personal communication). As an example, for one POTW in the MSDGC system the influent Cd loading was reduced by 57.5% over the period of 1971 through 1977 (Table 9) and for another plant in the system by 69.4% over the period of 1971 through 1984 (Table 10). Similar results were obtained with pretreatment in Baltimore (Table 11) and Philadelphia (Table 12). Once the federal pretreatment program has been fully implemented, reductions on the national level can be expected to approximate those achieved by these programs.

Table 11. Response of metals concentrations in digested sludge filter cake at the Back River POTW, Baltimore, Maryland in response to pretreatment efforts.[*]

Year	Cd	Cu	Pb	Ni	Zn
			mg/kg dry weight		
1978	51	2750	680	423	5000
1979	23	2540	539	397	3540
1980	18	2840	433	381	3400
1981	19	2070	493	374	3410
1982	18	1110	398	193	2360
1983	23	1060	324	214	2620
1984	26	1010	372	266	2750
1985	22	681	346	126	2030

[*]Source identification began in 1980, and source reduction began in 1981. Based on monthly composites in early years, then biweekly and weekly. Spencer, E. (1985, Personal communication).

Table 12. Response of metals concentrations in sludges at two Philadelphia POTWs in response to pretreatment program.[*]

Year	Cd	Cu	Pb	Ni	Zn
			mg/kg dry weight		
			Southwest		
1974	31	825	1540	100	3043
1976	27	1110	2710	103	2650
1977	27	1400	2170	185	3940
1978	16	1020	1800	275	4050
1980	18	986	740	98	2780
1981	25	971	562	117	2300
1982	20	940	1030	113	2440
1983	12.5	736	421	79	1700
1984	14.3	1140	427	111	1830
1985	15.0	880	373	80	1730
			Northeast		
1974	108	1610	2270	391	5391
1976	97	2240	2570	372	5070
1977	71	2320	2680	459	3920
1978	57	1240	1620	319	5910
1980	26	1210	728	275	3890
1982	14	985	423	185	2570
1983	10.9	1020	351	130	2110
1984	12.4	1200	360	130	1980
1985	17.3	1270	382	187	2100

[*]Source identification began in 1976. Liquid sludge analyzed until 1982, and sludge filter cake in 1983 and later. Semske, F. (1985, Personal communication).

Forms of Metals in Raw Sewage

The solubility of a metal in the soil-sludge mixture is inherently governed by the particular chemical form in which it occurs in the sludge. To understand the chemical form of the various metals in municipal sewage sludge, one must first determine what forms of these metals are affected by the wastewater treatment process. Each metal will be distributed between the soluble and solid phase based upon a complex equilibrium controlled by the wastewater composition. However, in most cases, Cd, Cr, Cu, Pb, and Zn have been found to be predominantly associated with the solid phase in wastewater influents.

Elenbogen et al. (1984) compared the raw sewage metals concentration entering a pilot-scale primary settling tank with the metals concentration of activated sludge solids (mixed liquor) in an activated sludge pilot plant. Elenbogen et al. (1984) (Table 13) concluded that the metals in raw sewage entering primary settling tanks are bound to the wastewater solids in similar proportions to the metals bound to the mixed liquor solids of the activated sludge process, and that the distribution of metals between the solid and liquid phases is the same.

Table 13. Cadmium uptake of mixed liquor sewage sludge (MLSS) at varying solids to Cd ratios.*

Run	Cd added†	Time of aeration	MLSS	MLSS to Cd ratio	Soluble Cd remaining	Cd uptake
	mg/L	min	mg/L		mg/L	%
A	1	0	1900	1900:1	0.0153	98.5
		30	1900	1900:1	0.0189	98.1
		60	1900	1900:1	0.0138	98.6
B	2	0	1000	500:1	0.375	81.2
		15	1000	500:1	0.186	90.8
		60	1000	500:1	0.122	93.9
		120	1000	500:1	0.112	94.6
		180	1000	500:1	0.066	96.7
C	10	0	9680	968:1	0.49	95.1
		15	9680	968:1	0.31	96.9
		30	9680	968:1	0.15	98.5
		60	9680	968:1	0.14	98.6
D	30	0	2600	87:1	7.50	75.0
		60	2600	87:1	2.28	92.4
		120	2600	87:1	1.82	94.0
		960	2600	87:1	0.78	97.4

*From Elenbogen et al. (1984).

†Added in the form of $CdCl_2$.

Metal adsorption by sludges has been demonstrated by spiking the sludge with inorganic metal salts. Elenbogen et al. (1984) studied the adsorptive capacity of the activated sludge process for metals and the strength of the bond between metals and the activated sludge solids. In batch-scale experiments in Table 13, Elenbogen et al. (1984) "spiked" mixed liquor with known amounts of $CdCl_2$ up to 30 mg/L. Most of the soluble Cd was adsorbed to the mixed liquor solids in less than 15 minutes and over 90% was adsorbed after 1 hour of aeration. In a separate series of controlled pH batch experiments, Elenbogen et al. (1984) also

studied the uptake of Cd by mixed liquor spiked with soluble Cd
($CdCl_2$) concentrations of 2.3 and 5.0 mg/L, and found that uptake
was not influenced by mixed liquor pH in the range of 5.0 to 8.0.
Over 90% of the soluble Cd uptake occurred in 15 minutes for all
of the pH levels tested.

Neufeld and Hermann (1975) also reported high adsorptive
capacities in batch-activated sludge reactors dosed with 30, 100,
and 300 mg/L soluble Cd. In their experiments, 65 to 70% of the
added soluble Cd was adsorbed on the activated sludge floc within
1 hour after dosing, even at the 300 mg/L soluble Cd dose.
Within 4 hours, 80% of the initial soluble Cd had been adsorbed
on the solids.

Patterson (1979) reported that in Rockford, Illinois, the
soluble fractions of Cd, Cr, and Zn were 24.0, 26.4, and 16.1%,
respectively, of these total metals in the raw sewage. Patterson
and Kodukula (1984) reported that in the raw sewage at a sewage
treatment facility of the MSDGC, Cd was 12.9% soluble, Cr was
1.7% soluble, Cu was 5.0% soluble, Pb was 16.9% soluble, Ni was
28.3% soluble, and Zn was 12.1% soluble. Similarly, Lester et
al. (1979) reported that 72% of the Cd, 70% of the Cu, and 73% of
the Pb was associated with the primary settled solids at the
Oxford, England treatment plant. In a companion study, Stoveland
et al. (1979) reported that 73% of the Cr and 74% of the Zn was
associated with the primary solids at the Oxford plant.

In general, therefore, studies of raw sewage metal speciation
indicate that most metals are associated with the solid phase.
This, coupled with the demonstrated affinity for soluble metals
shown by the activated sludge process, would indicate that most
of the metals contained in municipal sludge are associated with
the solid phase rather than the liquid phase.

Forms of Metals in Sludges

Because the chemical composition of municipal raw sewage and
the types of metal compounds that may enter a wastewater treatment plant vary widely the chemical transformations that will
occur in the plant are difficult to predict. However, a general
understanding of aqueous metal chemistry would suggest that
metals would be present in both organic and inorganic forms.
Metals associated with organic matter are probably bound strongly
to complexing sites. Inorganic forms could include metallic particles, relatively pure precipitates (phosphates. carbonates,
sulfides, or silicates), solid solutions resulting from coprecipitation with precipitates of Fe, Al, or Ca, or as metal ions
strongly adsorbed on surfaces of Fe, Al, or Ca minerals (Corey,
1981). If metals such as Cd or Zn enter the treatment plant in
aqueous form, coprecipitation with phosphates, hydrous oxides,
or sulfides of Fe and Al, and with phosphates and carbonates
of Ca would be expected (CAST, 1980; Logan and Chaney, 1983).

The scientific literature contains little specific analytical
data on the various species or forms of metals contained in
municipal sewage sludge. Investigators have focused on a
determination of sludge metal forms through the use of various

extractants. The amount of metal found in these various extractants is an indicator of the form of the metal in the sludge.

Stover et al. (1976) developed a sequential extraction scheme for fractionating Cd, Cu, Pb, Ni, and Zn in anaerobically digested sludge. In their scheme they suggest that KNO_3 extracts exchangeable metals, KF extracts adsorbed metals, $Na_4P_2O_7$ extracts organically bound metals, EDTA extracts metal carbonates and HNO_3 extracts metal sulfides. For the 12 sludges they studied, Stover et al. (1976) found that Zn was predominantly found in the organically bound ($Na_4P_2O_7$) form, Cu in the sulfide (HNO_3) form, and Pb in the metal carbonate (EDTA) form. Nickel was distributed in many forms, and Cd was predominantly in the metal carbonate (EDTA) form.

A similar extraction procedure, incorporating 0.5M KNO_3, "ion-exchange water", 0.5M NaOH, 0.05M Na_2EDTA, and 4.0M HNO_3 has been employed to fractionate Cd, Cu, Ni, and Zn in anaerobically digested air-dried sludge into forms designated as exchangeable, adsorbed, organically bound, carbonate, and sulfide/residual, respectively (Emmerich et al., 1982). While the Cd, Ni, and Zn occurred in sludge predominantly in carbonate form, the major forms of Cu extracted were in the order: organically bound > carbonate > sulfide/residual.

Six types of sludge from the MSDGC's West-Southwest Sewage Treatment Works were subjected to a sequential chemical extraction procedure in an effort to characterize the metal forms present in the sludges (Elenbogen, et al., 1983). The following conclusions were reached:

1. Lagoon sludge, waste-activated sludge, and filter cake had similar chemical distributions of Cd, Cu, and Zn with the predominant species (48 to 69%) of these metals being in the water soluble and readily exchangeable (KNO_3 extractable) forms. These values seem very high compared with those of other studies.)

2. Digested sludge had the highest percentage (22.5 to 25.4%) of Cd and Zn in the sulfide form ($\underline{1M}$ HNO_3) compared to the other sludges.

3. Heat-dried sludge and Nu Earth (air-dried MSDGC sludge) had similar chemical distributions for Cd and Zn, with a relatively small percentage (less than 10%) of these metals found in the water soluble form when compared to the other sludges (21 to 56%), and the greatest amount (34 to 75%) of Cd and Zn being recovered in the organically bound form ($Na_4P_2O_7$ extractable) for Nu Earth. However, in the case of heat-dried sludge, the greatest amount (29%) of Cu was found in the highly insoluble (concentrated HNO_3 extractable) form, with considerably less (17%) Cu being found in the organically bound form ($Na_4P_2O_7$ extractable) compared to the Nu Earth.

Metal Speciation in Soils

Metals in soils may be present in many forms. The application of sewage sludge to soils may alter the speciation of a metal, which, in turn, may affect its availability to plants.

The use of chemical extractants in studying metal speciation in soils has been focused mainly on the so-called plant available forms. Metals have frequently been extracted with simple aqueous solutions to determine plant available forms (Adams, 1965; Gupta and MacKay, 1966). In all cases, metal concentrations in water extracts were low.

Sequential chemical extraction schemes, considered to be of greater value than single extractants in determining metal distribution in wastewater sludge (Stover et al., 1976), have frequently been applied to fractionate trace metals in sludge-amended soils. A modified version of a sequential extraction procedure developed by Stover et al. (1976) was used by Emmerich et al. (1982) to determine the chemical forms of metals in loamy soils amended with anaerobically digested sludge. Emmerich et al. (1982) observed that less than 3% of the total Cd, Cu, Ni, and Zn in a sludge-amended loam soil were extracted by 0.5M KNO_3 and "ion-exchange water". Sposito et al. (1982) extracted 1.1 to 3.7% of these same metals using the same extractants. These results are consistent with exchangeable-plus-adsorbed forms of Cd, Pb, and Zn in sludges (Stover et al., 1976), and with water-soluble plus-exchangeable forms of Cd, Cu, Pb, and Zn in silt loam soils amended with digested sludge (Silviera and Sommers, 1977).

The diversity of reagents used to extract specific metal forms in soils make comparison of such studies difficult. Even if the reagent used is the same, the rate of leaching will be a function of the sample size, duration of extraction, temperature, and other factors (Sterritt and Lester, 1984).

Speciation of the metals in soils which receive sludge application is also important, as it will determine availability (Sterritt and Lester, 1984). However, Lake et al. (1984) conclude that no comprehensive or reliable speciation schemes for determining discrete heavy metal species or groups in sewage sludge and soil-sludge mixtures has yet been developed.

Plant-Availability of Sludge-Borne Trace Elements

The rate at which an element is taken up by a plant root appears to be a function of the activity of the free ion at the root surface (Checkai et al., 1982; Baker et al., 1984). However, as the concentration at the root surface decreases because of uptake, transport to the root may limit the rate of uptake.

The interacting factors that determine the rate of element uptake are most easily presented in a mathematical model. As most trace elements, particularly at low loadings, are delivered to the root surface primarily by diffusion (Barber, 1984), a diffusion model is used for illustrative purposes.

Factors Controlling Trace Element Uptake--
Theoretical Considerations

Soil factors that affect diffusive transport of a solute
include water content, solute concentration in solution, and the
ability to resupply absorbed solute (buffer power). . Important
plant factors include root geometry (root radius, presence of
root hairs/mycorrhizae) and root uptake physiology, i.e., root
absorbing power and effects of root exudates. How these factors
interact is shown in Eq. (1), which is a modification of an
uptake equation derived by Baldwin et al. (1973) that describes
the diffusive radial flux of solute from an isotropic medium
(soil) to a cylindrical sink (plant root), assuming depletion of
a cylindrical volume of soil surrounding each segment of root.

$$U = C_{1i}b \left\{ [1 - \exp] \left| \frac{-2\pi\alpha A_1 r_o L_v t}{b(1 + \frac{\alpha A_1 r_o}{D_1 \theta f} \ln \frac{r_h}{1.65 r_o})} \right| \right\} \quad (1)$$

Soil factors:

C_{1i} = initial concentration (mol/cm^3) of nutrient in soil
solution

b   = buffer power--the change in concentration of total
labile form [adsorbed + dissolved] (mol/cm^3 soil) per
unit of change in concentration of dissolved form
(mol/cm^3 soil solution)

A_1 = fractional area of soil solution (cm^2 water/cm^2 soil)

θ = volumetric water content (cm^3 water/cm^3 soil)

D_1 = diffusion coefficient in soil solution (cm^2/sec)

Plant factors:

U   = uptake per unit volume of soil in time, t (mol/cm^3 soil)

α = root absorbing power (uptake flux density, mol/cm^2
root·sec)/(concentration, mol/cm^3 soil solution)

t   = time (sec)

r_h = half-distance between roots (cm)

r_o = root radius (cm)

L_v = root density (cm root/cm^3 soil)

f   = conductivity factor (cm^2 soil/cm^2 water)

π = 3.1416

The conductivity factor, f, decreases with a decrease in θ because of greater tortuosity of the diffusion path at lower water contents. The buffer power, b, is equal to the change in concentration of total labile solute per unit change in concentration of that solute dissolved in the soil solution. The labile form includes dissolved and reactive adsorbed forms. Soils with high adsorption capacities for specific solutes generally show high buffer powers for those solutes. Commonly found ranges in buffer power for specific nutrients range from less than 1 for nonadsorbed species to more than 1000 for strongly adsorbed species, and generally decrease with increasing saturation of the adsorbing sites with a particular solute (Nye and Tinker, 1977; Barber, 1984).

The root density, L_v, is equal to the length of root per unit volume of soil. The value of L_v is readily determined for roots without root hairs or mycorrhizae, but their presence makes the geometry of the nutrient absorbing system more difficult to describe quantitatively. The root absorbing power, α, is equal to the uptake flux density divided by the nutrient concentration at the root surface. In some cases, α can be described by a Michaelis-Menten plot of flux density (mole per cm^2 per sec) vs. concentration at the root surface (mol/cm^3). This relationship has been shown to depend on the pre-existing nutrient status of the plant (Lauchli, 1984), speciation of dissolved solute (Checkai et al., 1982), and antagonistic effects of other metals (Logan and Chaney, 1983). A confounding factor is the effect of root exudates on rhizosphere pH (Marschner et al., 1982) and possible complexing of trace metals.

Sludge applications affect both C_{1i} and b in a soil. As the sludge application rate increases to the point where soil adsorption sites that can be filled at the activity supported by the pure sludge are nearly saturated, further increase in sludge application results in little additional change in either C_{1i} or b, and U should approach a maximum. The maximum uptake rate obtained with a given sludge should differ from soil-to-soil because pH differences affect both C_{1i} and b, and θ and f vary. Similarly, uptake will vary with plant species because of differences in α, L_v, r_o, and the nature of root exudates.

Most trace elements, particularly trace metals, added to soils appear to be immobilized mainly by adsorption reactions, which can usually be described by a Langmuir or Freundlich adsorption isotherm (Cavallaro and McBride, 1978; Garcia-Miragaya and Page, 1978; McBride et al., 1981; Kotuba, 1985). If the solute adsorption curve (adsorbed concentration vs. concentration in solution) for the soil and the desorption curve for the sludge have been determined, the equilibrium solute activity for any mixture of sludge and soil can be calculated. For example, if both curves can be described mathematically, in this case, by a Langmuir equation, the variables C_{1i} and b in the uptake equation can be calculated in the following way for a trace metal:

$$M = \frac{M_T C}{K + C} \quad \text{or} \quad C = \frac{M K}{M_T - M} \tag{2}$$

where M is metal adsorbed, M_T is the metal adsorption capacity, K is a constant equal to the dissolved metal concentration at one-half saturation of the adsorption sites, and C is the equilibrium metal activity. If a given amount of sludge is added to a given amount of soil, an amount of adsorbed metal, X, will be transferred from the sludge to the soil (if the sludge supports a higher metal activity than the soil), and a new equilibrium metal activity, C_M will result. If the amount of metal in solution is insignificant compared with that adsorbed, the reaction can be represented by the following equation, where the subscripts A and B represent soil and sludge, respectively.

$$C_M = \frac{(M_A + X)K_A}{M_{TA} - (M_A + X)} = \frac{(M_B - X)K_B}{M_{TB} - (M_B - X)} \tag{3}$$

At equilibrium, equation (3) can be used to solve for X and C_M. If X is small compared with M_B, the equilibrium metal activity will be very close to that for the sludge alone. In theory this approach would permit estimation of both metal solution activity and buffer power for input into an uptake model. The fact that sludge properties change with time presents some difficulty in practice.

Experimental Results

Few, if any, investigators have evaluated in a single study the interactions among plant uptake and sludge properties, sludge rate, and metal activity in the soil solution. Fujii (1983) found that application of a sludge containing 180 mg Cd/kg on sand and silt loam soils at pH 6 ± 0.3 maintained Cd activities in the sand 2.5 to 4 times those in the silt loam at rates up to 18.4 kg Cd/ha. Concentrations of Cd in corn tissue grown on these soils in the greenhouse were 1.5 to 2 times higher in the sand than the silt loam (Shammas, 1978). Adsorption characteristics of the sludge and soil were not studied.

Chelating resins have been used for determining Cd-adsorption characteristics of a muck and a sandy soil (Turner et al., 1984), and for characterizing dissolved metal complexes (Hendrickson et al., 1982; Hendrickson and Corey, 1983). Because of the large metal-ion buffering power of these resins, metal ions can be adsorbed on or desorbed from soils or sludges without significantly changing the metal-ion activities supported by the resin, if the proper ratio of resin to soil or sludge is used. Adaptation of this chelating-resin methodology to the routine determination of metal adsorption/desorption characteristics of soils and sludges appears promising.

Factors affecting the lability of metals in sludges have not been determined directly, but rather inferred from theoretical considerations, fractionation studies, or from greenhouse or field experiments with sludges of different chemical compositions. Greenhouse and field studies have generally supported the hypothesis (Corey, 1981; Corey et al., 1981) that much of the immobilization of trace metals in sludges is caused by coprecipitation with Fe, Al, and Ca precipitates during the treatment process, but no way of quantifying this effect has yet been devised.

In comparing the Cd uptake from equal rates of two noncalcareous sludges in the greenhouse, Cunningham et al. (1975) found that the average concentration of Cd in plant tissue was about the same for the 2 sludges (1.5 vs. 1.4 mg/kg), even though the Cd content of the sludges differed by a factor of 3 (76 vs. 220 mg/kg). The sludge with the lower Cd content had a lower Fe content (1.2 vs. 7.9%) and also a lower P content (2.9 vs. 6.1%). Thus, the effect of the higher Cd content in the one sludge may have been offset by a relatively high content of substances such as $FePO_4$ with which Cd could coprecipitate. In a field study, the Cd concentration in corn leaves from plots treated with a sludge containing 229 mg Cd/kg, 3.0% Fe, 1.1% Al, 4.7% Ca, and 1.6% P was nearly 3 times as high (1.7 vs. 0.6 mg/kg) as in corn leaves from plots treated with the same amount of Cd supplied by a sludge containing 180 mg Cd/kg, 7.8% Fe, 2.5% Al, 1.5% Ca, and 3.0% P (Keeney et al., 1980). The isotopically exchangeable Cd was also found to be 3 times higher for the sludge low in Fe and P, even though the total Cd concentrations in the 2 sludges were similar.

In a greenhouse study, Bates et al. (1979, Personal communication) added sludges to soils cropped to annual ryegrass over a period of about 5 years. Fourteen successive crops of ryegrass were grown, with sludge being added prior to seeding each crop. The cumulative Cd loadings were 10.6 kg/ha for the Sarnia sludge and 12.1 kg/ha for the Guelph sludge. The sludges had similar ratios of P to Cd at the start of the fourteenth crop, but the ratios of Fe to Cd were 889 and 195 for the Sarnia and Guelph sludges, respectively. The average Cd concentrations in the fourteenth crop of ryegrass were 1.35 mg/kg for the Sarnia sludge and 2.35 mg/kg for the Guelph sludge. With nearly equal additions of total Cd, the lower Cd availability was associated with the sludge having the higher Fe content. In fact, there was a measurable, though not statistically significant, decrease in plant-Cd concentration compared to the control treatment with a sludge in which the Fe and P contents were 8 and 5%, respectively, even though the total sludge-applied Cd was 1.63 kg/ha (Bates, 1986, Personal commununication).

Bell, et al. (1985, Personal communication) added 2 sludges with equal Cd but different Fe concentrations (sludges A and B in Table 14) to a fine sandy loam at rates high enough to show maximum Cd concentration in tobacco. The sludge with higher Fe showed lower Cd uptake even though the pH was slightly lower with that sludge.

Table 14. Effect of sludge properties on plateau C concentrations of
Cd in tobacco leaves grown in the field long after sludge
application.*

Sludge	Concentration in sludge Cd	Concentration in sludge Fe	Maximum application		Soil pH	Increased Cd above control at plateau
	mg/kg	%	mt/ha	kg Cd/ha		mg Cd/kg dry wt.
A	13.2	2.5	224	2.90	5.4	7.5
B	13.4	8.3	224	3.00	5.2	2.4
	13.4	8.3	224	3.00	5.8	0.2

*Bell, et al. (1985, Personal communication).

†Sludge A applied in 1972 to Beltsville silt loam. Sludge B applied
in 1976 to Christiana fine sandy loam. Tobacco ('Maryland 609')
grown in 1983 and 1984 for Sludge A, and 1984 for Sludge B.

Additional evidence that metal concentration in plants may
be affected by the form of metals in sludge can be implied from
the work of King and Dunlop (1982). Sludges from Wilmington,
North Carolina (13 mg Cd/kg) and Philadelphia, Pennsylvania
(225 mg Cd/kg) were applied to several soils in which corn was
grown in the greenhouse. Sludges were applied at rates to supply
equal amounts of Cd. The effect of sludge type was significant,
as evidenced by different slopes in models of Cd concentration
in corn stover regressed on Cd loading rate:

Wilmington: plant Cd (mg/kg) = 0.11 + 0.18 Cd rate (kg/ha)
Philadelphia: plant Cd (mg/kg) = 0.18 + 0.45 Cd rate (kg/ha)

The relationship between sludge rate and metal uptake by
plants has been investigated by many researchers. In general,
for Cd and Zn the metal concentration in tissue approaches a
maximum and has shown a logarithmic or Langmuir-type relationship
with sludge rate. In many cases, Cu uptake is not affected
significantly. In studies when the sludge addition was not high
enough to approach a constant metal activity in solution, the
relationship between sludge rate and tissue concentration often
approached linearity. For example, Pietz et al. (1983) and
Hinesly et al. (1984) found a near linear relationship between
Cd or Zn in corn leaves and Cd application rate up to 111 kg/ha
applied to a calcareous strip mine spoil in a sludge containing
about 300 mg Cd/kg. The sludge was applied over a period of 6
years. When the same sludge was applied for 12 years to an acid
Blount soil, the year-to-year variability in leaf analyses
obscured any relationship to cumulative additions (Hinesly et
al., 1984). Vlamis et al. (1985) also noted a linear response
with Cd and Zn in barley with 2 sludges applied at rates up to
225 mt/ha on an acid soil. However, metal uptake at the high
sludge applications may have been augmented by the effects of
lower pH found in these treatments.

In contrast to the linear responses reported above, Soon et al. (1980) found that Cd in corn stover was logarithmically related to sludge rate (up to 4 kg Cd/ha) for 3 sludges receiving either Fe, Al, or Ca additions during treatment. All relationships were logarithmic, but the plots of the Cd concentration in stover and the log of Cd applied differed in slope by a factor of 3, emphasizing the effects of sludge properties on Cd availability.

In a field study, Chaney et al. (1982) used sludges with different Cd concentrations to determine the effect of concentration and application rate on Cd concentration in lettuce. The first increment of low-Cd sludge (13 mg/kg) increased Cd concentration in the lettuce slightly, but higher rates had no further effect on Cd concentration. The first increment of high-Cd sludge (210 mg/kg) had a pronounced effect on Cd concentration, but the response to additional increments was less pronounced, indicating a logarithmic response (Figure 2). Unpublished data on Cd accumulation in lettuce (Chaney, 1985, Personal communication) (Figure 3) also show a plateau effect along with an effect of pH and sludge composition.

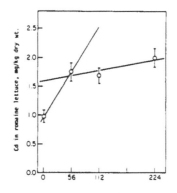

Sludge application rate (mt/ha)

Figure 2. Uptake of Cd by romaine lettuce from soils treated with municipal sewage sludge at various rates (mean data over 8 years).

In another field study (Bell et al., 1985, Personal communication), tobacco was grown on sludge-treated soil. Copper, Cd, and Zn contents of the tobacco were affected markedly by the first increment of sludge. Additional increments had no effect on Cu content, and only slightly increased Zn and Cd content.

Sommers (1985, Personal communication) presented data in which 3 sludges containing high concentrations of Cd (284, 1210, and 247 mg Cd/kg) were applied to a Chalmers soil at linearly increasing rates for 2 sludges, and logarithmically increasing

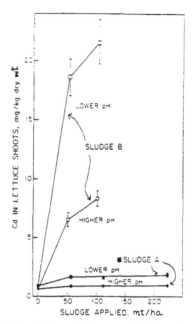

Figure 3. Effect of sludge application rate on Cd in lettuce leaves.
Sludge A (13.4 mg Cd/kg) and 8.3% Fe) applied in 1976 and
lettuce grown in 1976 to 1983; sludge B (210 mg Cd/kg) and
2.5% Fe) applied in 1978 and lettuce grown in 1978 to 1983.
Results shown are geometric means over years ± standard
error (Chaney, 1985, Personal communication).

rates for the third (Table 15). Over a period of 8 years, oats,
winter wheat, soybeans, and corn were grown on these plots. In
almost all cases, the relationship between tissue concentration
of Cd or Zn and rate of metal applied in the form of sludge was
logarithmic or approached a constant value at high sludge rates
(Table 15). This was also the case with vegetable data from the
Metropolitan Sanitary District of Greater Chicago (Table 16) and
of Hinesly et al. (1984). Crops studied by Hinesly et al. (1984)
included corn (Figure 4) and many other grass species (Table 17)
Hinesly and Redborg (1984) grown for 3 years following sludge
application. One interesting aspect of this latter study was
the decrease in Cd concentration in corn tissue with time after
application, particularly at the high rate (Figure 4). This
effect is further illustrated in data from Dowdy et al. (1984)
which show uptake by corn silage in grams per hectare over a
period from 1979-1984 (Figure 5). These studies agree with the
conclusions of CAST (1980) and Logan and Chaney (1983), that
bioavailability of metals remains constant or decreases over a
period of years at a constant pH. The marked decrease in metal
uptake at the high rate after 1 year suggests that exposure of
anaerobically-digested sludges to an aerobic environment, and/or
interactions with soil may produce marked changes in lability
of the metals during the first year following application.

Table 15. Effect of sludge rate and year after sludge application on concentrations of cadmium, zinc, copper, and nickel in oat straw and leaves of winter wheat, soybean, and corn.[*]

Sludge	Metal	Oats (1)	Winter wheat (2)	(8)	Soybean (1)	(8)	Corn (1)	(8)[‡]
	kg/ha				mg/kg			
				Cadmium				
A	0	0.9	0.5	0.3	1.6	1.4	1.3	0.6
	16	1.3	1.5	0.3	2.7	1.8	1.6	1.2
	32	2.0	2.0	0.4	1.8	1.9	1.5	1.0
	64	2.2	5.4	0.5	1.8	3.4	1.6	1.0
	127	3.3	6.5	0.5	2.4	4.9	1.9	1.1
FK	0	0.9	0.5	0.3	1.6	1.4	1.3	0.6
	68	9.9	12.1	1.2	4.6	5.0	5.1	5.6
	136	15.9	14.6	1.4	5.0	8.9	7.8	4.7
	203	20.3	15.8	1.5	6.0	11.0	—	—
MA	0	0.9	0.5	0.3	1.6	1.4	1.3	0.6
	14	0.9	1.1	0.3	2.1	1.2	1.1	1.1
	28	1.1	1.0	0.3	1.7	1.3	1.4	1.1
	42	1.1	1.4	0.3	2.1	2.4	0.9	0.8
				Zinc				
A	0	20.2	21.9	35.3	41.6	72.2	25.7	69.8
	381	30.3	56.3	38.1	53.0	103.5	39.1	62.7
	767	41.4	67.4	44.2	66.8	92.6	53.7	71.3
	1523	48.9	63.5	41.4	55.0	110.0	48.7	77.0
FK	0	20.2	21.9	35.3	41.6	72.2	25.7	69.8
	106	27.0	52.8	40.1	54.4	79.1	50.5	55.2
	213	30.3	50.5	36.9	59.1	80.1	48.5	54.1
	319	32.0	48.8	40.1	61.2	82.9	—	—
MA	0	20.2	21.9	35.3	41.6	72.2	25.7	69.8
	291	20.2	38.0	32.8	44.1	76.7	44.4	78.1
	582	26.0	33.1	38.5	48.4	88.5	44.7	67.8
	875	22.6	30.3	35.6	51.2	86.7	52.7	83.9
				Copper				
A	0	2.0	4.0	2.6	7.4	9.1	8.3	5.5
	67	2.2	4.1	2.5	8.7	9.4	8.2	5.9
	134	3.0	5.0	3.4	8.1	9.4	9.2	6.4
	269	4.9	11.0	4.0	7.7	9.7	9.4	7.5
	538	6.0	7.8	4.8	7.4	10.2	10.3	7.7
FK	0	2.0	4.0	2.6	7.4	9.1	8.3	5.5
	74	2.6	4.6	4.8	9.8	10.1	11.4	6.7
	149	4.2	4.3	3.7	10.1	10.6	10.6	6.8
	223	3.7	4.5	4.7	10.3	11.2	—	—
A	0	2.0	4.0	2.6	7.4	9.1	8.3	5.5
	25	2.5	3.7	3.2	8.2	10.0	7.9	7.1
	50	2.5	4.9	3.1	8.4	10.9	8.9	6.7
	76	2.6	4.0	2.7	10.0	10.2	9.6	4.7
				Nickel				
A	0	1.1	2.9	0.3	49.3	5.7	3.1	0.2
	114	2.0	3.2	0.3	18.5	9.5	3.9	0.3
	228	3.3	4.0	0.4	20.2	9.3	1.6	0.2
	451	5.1	9.8	0.4	28.5	10.9	5.9	0.3
	914	12.5	7.5	0.8	20.8	16.6	2.5	0.3
FK	0	1.1	2.9	0.3	49.3	5.7	3.1	0.2
	24	1.6	4.1	0.4	44.1	6.9	2.3	0.3
	48	2.5	2.8	0.6	28.4	8.6	2.6	0.2
	72	3.4	3.6	0.5	19.2	9.2	—	—
A	0	1.1	2.9	0.3	49.3	5.7	3.1	0.2
	12	1.1	2.5	0.4	21.3	3.7	1.3	0.5
	24	1.2	2.7	0.5	24.2	5.2	6.8	0.5
	36	1.3	2.7	0.4	11.4	4.2	1.4	0.4

[*]Sommers (1985, Personal communication).

[†]Leaf tissue for winter wheat, soybean and corn; straw for oats.

[‡]Numbers in parentheses denote crop year following sludge application.

Table 16.   Cadmium, zinc, copper, and nickel concentrations in edible
parts of vegetables grown at west-southwest sewage treatment
works, Metropolitan Sanitary District of Greater Chicago.[*]

Crop	Year	Sludge added, mg/ha				
		0	60	120	240	300
		— — — — — mg metal/kg edible tissue — — — —				
Cadmium							
Beets	1979	0.2	1.4	1.6	2.7	2.9
	1983	0.7	1.8	2.0	1.8	4.9
Tomatoes	1979	1.1	1.8	2.4	2.2	3.4
	1983	2.0	2.2	2.1	2.6	2.9
Swiss	1979	2.3	8.0	12.2	16.8	22.1
chard	1983	1.3	4.0	8.1	8.4	12.2
Carrots	1979	0.7	1.4	1.2	1.9	2.3
	1983	1.3	1.8	2.8	3.4	3.1
Green	1979	0.4	0.3	0.4	0.4	0.5
beans	1983	0.1	0.1	0.2	0.1	0.2
Spinach	1979	6.4	12.6	10.3	14.4	12.1
	1983	8.5	21.4	28.3	31.7	33.8
	(1982)	10.3	11.3	17.2	10.6	11.8
Zinc							
Beets	1979	34	45	62	93	90
	1983	39	55	59	80	97
Tomatoes	1979	27	30	30	34	35
	1983	27	32	31	29	32
Swiss	1979	69	129	176	237	302
chard	1983	62	91	120	129	251
Carrots	1979	20	22	25	27	32
	1983	31	33	30	31	33
Green	1979	32	33	37	37	37
beans	1983	38	35	33	35	35
Spinach	1979	147	249	265	309	311
	1983	209	433	404	451	472
	(1982)	201	276	315	226	258
Copper							
Beets	1979	8.8	10.4	9.8	11.1	12.8
	1983	9.7	11.9	12.3	14.2	14.1
Tomatoes	1979	13.2	16.6	14.5	16.2	16.6
	1983	10.4	12.4	11.9	11.0	12.0
Swiss	1979	23.7	20.8	25.6	30.9	29.2
chard	1983	25.2	26.6	27.6	26.7	29.4
Carrots	1979	5.4	6.0	5.8	5.9	6.5
	1983	7.9	7.8	7.5	7.8	7.7
Green	1979	8.6	8.9	9.2	8.2	8.5
beans	1983	8.5	9.1	7.6	8.6	7.9
Spinach	1979	13.8	14.2	17.3	18.9	19.1
	1983	12.3	14.2	15.1	16.3	17.4
Nickel							
Beets	1979	0.5	0.6	0.9	1.4	1.5
	1983	2.1	2.6	2.8	3.3	6.6
Tomatoes	1979	1.1	1.3	1.4	4.1	2.7
	1983	6.8	7.1	27.1	7.2	8.4
Swiss	1979	1.3	1.4	2.5	3.5	4.3
chard	1983	0.8	2.3	2.4	2.1	3.3
Carrots	1979	0.6	1.1	1.4	2.2	2.9
	1983	1.9	0.7	0.8	1.2	1.0
Green	1979	3.3	1.2	2.4	3.6	3.3
beans	1983	2.3	2.1	2.9	5.1	4.7
Spinach	1979	1.4	1.5	1.7	2.7	2.6
	1983	10.6	5.8	8.8	6.0	10.0

[*]Nu Earth was applied from 1977 through 1979 in three equal applications
(C. Lue-Hing, 1985, Personal communication).

In addition to metal forms and concentrations in sludges, the effects of sludge additions on soil pH must also be considered. Lime-stabilized sludges function as liming materials (Soon et al., 1980; Chang et al., 1980). Noncalcareous sludges may either raise or lower soil pH, depending on such factors as amount of NH_4-N nitrified or sludge pH, and pH buffering capacity in relation to the pH and buffering capacity of the soil. Generally, plant uptake of metals increases with increasing acidity (CAST, 1980; Logan and Chaney, 1983).

Table 17. Effect of sludge rate applied in 1979 on concentrations of cadmium, zinc, and copper in wheat, rye and four grasses in 1981.[*]

Crop	Sludge rate, mg/ha			
	0	224	448	896
	- - - - - mg metal/kg tissue - - - - - -			
Cadmium				
Wheat (leaf)	<0.1	0.7	1.6	1.5
Rye (leaf)	0.1	0.4	0.7	0.9
Redtop	0.1	0.3	0.8	1.2
Brome	0.1	0.8	1.0	1.6
Orchard grass	0.2	0.5	1.4	1.7
Western wheat grass	<0.1	0.4	0.8	0.7
Reed canary grass	0.1	0.2	0.6	0.8
Perennial rye	<0.1	0.1	0.9	0.7
Timothy	0.1	0.1	0.2	0.8
Tall fescue	0.1	0.4	0.9	1.8
Zinc				
Wheat (leaf)	15	26	39	46
Rye (leaf)	20	42	71	61
Redtop	19	45	68	86
Brome	17	29	40	48
Orchard grass	21	32	39	49
Wetern wheat grass	19	35	41	44
Reed canary grass	28	62	104	121
Perennial rye	17	40	90	88
Timothy	23	50	65	74
Tall fescue	23	35	48	62
Copper				
Wheat (leaf)	7	8	9	9
Rye (leaf)	10	13	18	18
Redtop	5	9	12	14
Brome	5	7	8	10
Orchard grass	4	6	7	9
Western wheat grass	4	7	8	8
Reed canary grass	6	9	11	13
Perennial rye	4	6	10	17
Timothy	5	9	10	12
Tall fescue	5	7	8	10

[*]Hinesly and Redborg (1984)

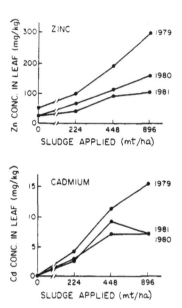

Figure 4. Effect of a one-time application of municipal sewage sludge containing 4230 mg Zn/kg and 300 mg Cd/kg on the Zn and Cd contents of corn leaf tissue in each of 3 years after application to a calcareous strip-mine spoil. Data for 1979 and 1980 are from Hinesly et al. (1984a). Data for 1981 are from Table 17.

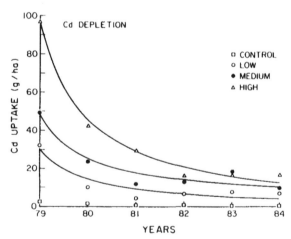

Figure 5. Decrease in Cd uptake by corn silage with time after application of sewage sludge at three rates in 1979 (Dowdy et al., 1984).

CONCLUSIONS

1. Trace metals in the influent to a sewage treatment plant are associated mainly with the solids, and they remain associated with solids in the sludge following treatment.

2. Concentrations of trace elements in many POTW sludges have decreased markedly in the past decade as a result of industrial waste pretreatment, and this trend is expected to continue.

3. Uptake of solutes by plants is influenced by soil and plant factors, and a simplified mathematical model is presented to indicate how these variables interact in affecting solute uptake.

4. The relevant soil variables related to solute uptake affected most by sludge application are concentration in solution, concentration of the labile adsorbed form, and distribution of dissolved species between free-ion and complexed forms.

5. The relevant plant variables related to solute uptake affected most by sludge application are root adsorbing power (related to speciation in solution and concentrations of other ions competing for uptake sites) and possibly root geometry.

6. The equilibrium trace element concentrations that a sludge supports depend on the chemical properties of the sludge, particularly the presence of trace-element precipitates, whether relatively pure or coprecipitated with Fe, Al, or Ca precipitates, the strength of bonding to organic and mineral adsorption sites, the proportion of potential adsorbing sites filled, and the presence of dissolved ligands capable of complexing the trace metals. If sludge matrix is constant, plant availability of a trace element increases with increasing concentration of that trace element in the sludge.

7. If a sludge supports a higher equilibrium solution concentration of an element than does a soil, mixing the 2 will result in an equilibrium concentration intermediate between the 2 that should be predictable if the desorption characteristics of the sludge and adsorption characteristics of the soil are known. This hypothesis has not been tested experimentally.

8. As increasing amounts of sludge are added to a soil, trace-element adsorption sites on the soil become progressively saturated (or desaturated) to the point that the equilibrium concentration approaches that of the sludge alone. Further sludge applications above a particular level depending on soil properties should result in little, if any, change in equilibrium concentration. Below this critical sludge rate,

soil adsorption characteristics affect the equilibrium concentration supported by a given addition of sludge. Above that critical sludge rate, the equilibrium concentration characteristic of the sludge should be maintained (sludge controls).

9. If the equilibrium trace-element concentration (and buffer power) supported by the sludge is less than that which will result in excessive concentrations in plant tissue or damage to the plant, there is no need to limit application rates of that sludge on the basis of metal content.

10. If the equilibrium metal concentration (and buffer power) supported by the sludge at a specified pH is high enough to cause excessive concentrations in plant tissue or plant damage, determining maximum loading rates based on both soil and sludge characteristics will be required.

11. Most research indicates that plant availability of sludge-derived metals stays the same or decreases with time following application. Therefore, any testing procedures developed to estabish long-term metal-loading limits should be run after the sludge has been allowed to equilibrate with the soil. Presently the time(s) required to equilibrate sludges with soils are not precisely known, but limited data suggest a minimum of two cropping seasons.

12. Methods involving chelating resins for obtaining metal desorption curves for sludges and metal adsorption curves for soils appear promising.

13. Addition of Fe or Al salts or lime during the sewage treatment process appears to reduce equilibrium metal activities supported by sludge; however, research designed to test this hypothesis has not been done.

REFERENCES

Adams, F. 1965. Manganese. In C. A. Black et al. (ed.) Methods of soil analysis. Part 2. Agronomy 69:1011-1018.

Baker, D. E., D. S. Rasmussen, and J. Kotuby. 1984. Trace metal interactions affecting soil loading capacities for cadmium. pp. 118-132. In L. P. Jackson et al. (ed.) Hazardous and Industrial Waste Management and Testing. Third ASTM Symposium, Philadelphia, PA. 7-10 March 1983. Special Technical Publication 851. Am. Soc. for Testing and Materials, Philadelphia, PA.

Baldwin, J. P., P. H. Nye, and P. B. Tinker. 1973. Uptake of solutes by multiple root systems from soil. III. A model for calculating the solute uptake by a randomly dispersed root system developing in a finite volume of soil. Plant Soil 38:621-635.

Barber, S. A. 1984. Soil nutrient bioavailability: A mechanistic approach. 1st. ed. John Wiley and Sons, NY.

Bates, T. E. 1986. Personal communication. Dept. of Land Resources Science, Univ. of Guelph, Ontario N1G 2W1 Canada.

Bates, T. E., E. Haq, and Y. K. Soon. 1979. Report. Dept. of Land Resources Science, University of Guelph, Ontario.

Bell, P. F., C. L. Mulchi, and R. L. Chaney. Personal communication, Univ. of Maryland, College Park, MD 20742.

Bell, P. F., C. L. Mulchi, and C. Adamu. 1985. Long-term availability of metals in sludge amended soils used for the growth of Nicotiana tabacum L. Agron. Abstr. 1985:21.

Council for Agricultural Science and Technology (CAST). 1980. Effects of sewage sludge on the cadmium and zinc content of crops. Counc. Agric. Sci. Tech. no. 83. Ames, IA.

Cavallaro, N., and M. B. McBride. 1978. Copper and cadmium adsorption characteristics of selected acid and calcareous soils. Soil Sci. Soc. Am. J. 42:550-556.

Checkai, R. T., R. B. Corey, and P. A. Helmke. 1982. The effects of ionic and complexed Cd on metal uptake by plants. Agron. Abstr. American Society of Agronomy, Madison, WI, p. 93.

Chaney, R. L. 1985. Personal communication. USDA-ARS-NER, Rm. 101, Bldg. 008, BARC-West, Beltsville, MD 20705.

Chaney, R. L., S. B. Sterrett, M. C. Morella, and C. A. Lloyd. 1982. Effect of sludge quality and rate, soil pH, and time on heavy metal residues in leafy vegetables. p. 444-458. In Proc. Fifth Annual Madison Conf. Appl. Res. Prac. Municipal Industrial Waste., Madison, WI. 22-24 Sept. 1982. Univ. of Wisconsin-Extension, Madison, WI.

Chang, A. C., A. L. Page, and F. T. Bingham. 1980. Re-utilization of municipal wastewater sludges—metals and nitrate. J. Water Pollut. Control Fed. 53:237-245.

Cheng, D. G., P. A. Snow, M. C. B. Fanning, and D. S. Fanning. 1985. Liming acid sulfate soils in Baltimore Harbor with dredged materials. Agron. Abstr. American Society of Agronomy, Madison, WI, p. 23.

Corey, R. B. 1981. Adsorption vs. precipitation. p. 161-182. In M. A. Anderson and A. J. Rubin (ed.) Adsorption of inorganics at solid-liquid interfaces. Ann Arbor Science Publ., Inc., Ann Arbor, MI.

Corey, R. B., R. Fujii, and L. L. Hendrickson. 1981. Bioavailability of heavy metals in soil-sludge systems. p. 449-465. In Proc. Fourth Annual Madison Conf. Appl. Res. Pract. Municipal Ind. Waste, Madison, WI. 28-30 Sept. 1981, Univ. of Wisconsin-Extension, Madison, WI.

Cunningham, J. D., D. R. Keeney and J. A. Ryan. 1975. Phytotoxicity and uptake of metals added to soils as inorganic salts or in sewage sludge. J. Environ. Qual. 4:460-461.

Dowdy, R. H., R. D. Goodrich, W. E. Larson, B. J. Bray, and D. E. Pamp. 1984. Effects of sewage sludge on corn silage and animal products. Report EPA-600-2-84-075, Municipal Environmental Research Laboratory, Office of Research and Development, U.S. Environmental Protection Agency, Cincinnati, OH.

Elenbogen, G., B. Sawyer, C. Lue-Hing, and D. R. Zenz. 1983. Sludge metal species as determined by their solubilities in different reagents - Phase I report. Res. and Develop. Dep. Rep. no. 83-31, The Metropolitan Sanitary District of Greater Chicago, Chicago, IL.

Elenbogen, G., B. Sawyer, K. C. Rao, D. R. Zenz, and C. Lue-Hing. 1984. Studies of the uptake of heavy metals by activated sludge. Res. and Develop. Dep. Rep. no. 84-7, The Metropolitan Sanitary District of Greater Chicago, Chicago, IL.

Emmerich, W. E., L. J. Lund, A. L. Page, and A. C. Chang. 1982. Solid phase forms of heavy metals in sewage sludge-treated soil. J. Environ. Qual. 11:178-181.

Fujii, R. 1983. Determination of trace metal speciation in soils and sludge-amended soils. Ph.D. Diss., Univ. of Wisconsin, Madison, Diss. Abstr. 83-15000.

Garcia-Miragaya, J., and A. L. Page. 1978. Sorption of trace quantities of cadmium by soils with different chemical and mineralogical composition. Water, Air, Soil Pollut. 9:289-299.

Gupta, U. C., and D. C. MacKay. 1966. The relationship of soil properties to exchangeable and water soluble copper and molybdenum status in podzol soils of Eastern Canada. Soil Sci. Soc. Am. Proc. 30:373-375.

Hendrickson, L. L., M. A. Turner, and R. B. Corey. 1982. Use of Chelex-100 to maintain constant metal activity and its application to characterization of metal complexation. Anal. Chem. 54:1633-1637.

Hendrickson, L. L. and R. B. Corey. 1983. A chelating-resin method for characterizing soluble metal complexes. Soil Sci. Soc. Am. J. 47:467-474.

Hinesly, T. D., K. E. Redborg, R. I. Pietz, and E. L. Ziegler. 1984. Cadmium and zinc uptake by corn (Zea mays L.) with repeated applications of sewage sludge. J. Agric. Food Chem. 32:155-163.

Hinesly, T. D., and K. E. Redborg. 1984. Long-term use of sewage sludge on agricultural and disturbed lands. EPA-600/2-84-126. Municipal Environmental Research Laboratory, Office of Research and Development, U.S. Environmental Protection Agency, Cincinnati, OH.

Keeney, D. R., R. B. Corey, P. A. Helmke, L. L. Hendrickson, R. L. Koretev, M. A. Turner, A. T. Shammas, K. D. Kunz, R. Fujii, and P. H. Williams. 1980. Heavy metal bioavailability in sludge-amended soils. U.S. EPA Project R8046140. Final Report. (Unpublished).

King, L. D., and W. R. Dunlop. 1982. Application of sewage sludge to soils high in organic matter. J. Environ. Qual. 11:608-616.

Kotuba, J. 1985. The availability of cadmium as determined by retention/release reactions, trace metal speciation, and plant uptake. M.S. Thesis, The Pennsylvania State Univ., University Park, PA.

Lake, D. L., P. W. W. Kirk, and J. N. Lester. 1984. Fractionation, characterization and speciation of heavy metals in sewage sludge and sludge-amended soils: A review. J. Environ. Qual. 13:175-183.

Läuchli, A. 1984. Mechanisms of nutrient fluxes at membranes of the root surface and their regulation in the whole plant. p. 1-25. In S. A. Barber and D. R. Bouldin (ed.) Roots, nutrient and water influx and plant growth. Spec. Pub. 49. American Society of Agronomy, Madison, WI.

Lester, J. N., R. M. Harrison, and R. Perry. 1979. The balance of heavy metals through a sewage treatment works. I. Lead, cadmium, and copper. The Science of the Total Environment 12:13-23.

Logan, T. J., and R. L. Chaney. 1983. Utilization of municipal wastewater and sludge on land—metals. p. 235-326. In A. L. Page, et al. (ed.) Proceedings of the workshop on utilization of municipal sludge on land. Univ. of California, Riverside, CA.

Lue-Hing, C. 1985. Personal communication. Metropolitan Sanitary District of Greater Chicago. 100 East Erie St., Chicago, IL 60611.

Marschner, H., V. Römheld, and H. Ossenberg-Neuhaus. 1982. Rapid method for measuring changes in pH and reducing processes along roots of intact plants. Z. Pflanzenphysiol. Bd. 105:407-416.

McBride, M. B., L. D. Tyler, and D. A. Hovde. 1981. Cadmium adsorption by soils and uptake by plants as affected by soil chemical properties. Soil Sci. Soc. Am. J. 45:739-744.

Neufeld, R. D., and E. R. Hermann. 1975. Heavy metal removal by acclimated activated sludge. J. Water Pollut. Control Fed. 47:310-329.

Nye, P. H., and P. B. Tinker. 1977. Solute movement in the soil-root system. Univ. of California Press, Berkeley and Los Angeles, CA.

Patterson, J. W. 1979. Parameters influencing metals removal in POTWs. Natl. Academy of Sciences Symp. on Management and Control of Toxic Materials in Municipal Sludges, Miami Beach, FL.

Patterson, J. W., and P. S. Kodukula. 1984. Metals distributions in activated sludge systems. J. Water Pollut. Control Fed. 56:432-441.

Pietz, R. I., J. R. Peterson, T. D. Hinesly, E. L. Ziegler, K. E. Redborg, and C. Lue-Hing. 1983. Sewage sludge application to calcareous strip-mine spoil: II. Effect on spoil and corn cadmium, copper, nickel and zinc. J. Environ. Qual. 12:463-467.

Semske, F. 1985. Personal communication. City of Philadelphia Water Dept., Municipal Service Bldg., 15th and JFK Blvd., Philadelphia, PA 19107.

Shammas, A. T. 1978. Bioavailability of cadmium in sewage sludge. Ph.D. Diss., Univ. of Wisconsin, Madison, WI. Diss. Abstr. 79-19813.

Silviera, D. J., and L. E. Sommers. 1977. Extractability of copper, zinc, cadmium, and lead in soils incubated with sewage sludge. J. Environ. Qual. 6:47-52.

Sommers, L. E. 1977. Chemical composition of sewage sludges and analysis of their potential use as fertilizers. J. Environ. Qual. 6:225-232.

Sommers, L. E. 1985. Personal communication. Dept. of Agronomy, Colorado State Univ., Fort Collins, CO 80523.

Soon, Y. K., T. E. Bates, and J. R. Moyer. 1980. Land application of chemically treated sewage sludge: III. Effects of soil and plant heavy metal content. J. Environ. Qual. 9:497-504.

Spencer, E. 1985. Personal communication. Maryland State Dept. of Health, Dept. of Solid Waste, 261 W. Preston St., Baltimore, MD 21201.

Sposito, G., F. T. Bingham, S. S. Yadav, and C. A. Inouye. 1982. Trace metal complexation by fulvic acid extracted from sewage sludge. II. Development of chemical models. Soil Sci. Soc. Am. J. 46:51-56.

Sterritt, R. M., and J. N. Lester. 1984. Significance and behavior of heavy metals in wastewater treatment processes. III. Speciation in wastewaters and related complex matrices. The Science of the Total Environment 34:117-141.

Stoveland, S., M. Astruc, J. N. Lester, and R. Perry. 1979. The balance of heavy metals through a sewage treatment works. II. Chromium, nickel, and zinc. The Science of the Total Environment 12:25-34.

Stover, R. C., L. E. Sommers, and D. J. Silviera. 1976. Evaluation of metals in wastewater sludge. J. Water Pollut. Control Fed. 48:2165-2175.

Turner, M. A., L. L. Hendrickson, and R. B. Corey. 1984. Use of chelating resins in metal adsorption studies. Soil Sci. Soc. Am. J. 48:763-769.

Vlamis, J., D. E. Williams, J. E. Corey, A. L. Page, and T. J. Ganje. 1985. Zinc and cadmium uptake by barley in field plots fertilized seven years with urban and suburban sludge. Soil Sci. 139:81-87.

Whitebloom, S. W., C. Lue-Hing, E. Guth, and J. Dencek. 1978. The development and enforcement of the industrial waste and control ordinance at the Metropolitan Sanitary District of Greater Chicago. Res. and Develop. Dep. Rep. no. 78-11, The Metropolitan Sanitary District of Greater Chicago.

CHAPTER 4

EFFECTS OF LONG-TERM SLUDGE APPLICATION ON
ACCUMULATION OF TRACE ELEMENTS BY CROPS

Andrew C. Chang
University of California, Riverside, California

Thomas D. Hinesly
University of Illinois, Urbana, Illinois

Thomas E. Bates
University of Guelph, Guelph, Ontario, Canada

Harvey E. Doner
University of California, Berkeley, California

Robert H. Dowdy
USDA-ARS, University of Minnesota, St. Paul, Minnesota

James A. Ryan
U.S. Environmental Protection Agency, Cincinnati, Ohio

INTRODUCTION

Since the last comprehensive review of elemental uptake by
plants grown on sludge treated soils (Logan and Chaney, 1983),
considerable data from long-term field experiments have become
available. Most experiments were designed to assess the effects
of sludge applications on plant accumulation of metals (e.g., Cd,
Cu, Ni, Zn, etc.). This chapter will concentrate on newly
available long-term field data in terms of their implications on
land application of sludges.
   In the following sections, we will attempt to answer:
   (1) What is the quality of experimental data?
   (2) Do repeated annual sludge applications affect the metal
       accumulation in plant tissue?
   (3) Do the plant tissue metal accumulation patterns of a
       single sludge application differ from those of multiple
       sludge applications having equal total metal input? and
   (4) Does metal uptake by plants change following termination
       of sludge application?

Nature of the Experimental Data

Logan and Chaney (1983) pointed out that common errors in the study of toxic element uptake by plants grown in sludge treated soils are (1) substituting inorganic metal salts for sludge or spiking the sludge with inorganic salts in preparing the growth medium; (2) relying on short term small pot experiments in the greenhouse rather than field observations to predict metal concentrations in plants. Recently data from field studies have become available which allows us to greatly reduce our reliance upon information flawed by the above mentioned errors. During the course of our deliberation, only in the absence of field data did we draw upon "large" pot, greenhouse or growth chamber findings. It has not been necessary to use any data derived from studies that used "salts" as the metal source.

The drawing of "general" conclusions from a pool of information derived from unrelated field studies have limitations that must be recognized. (For example, plant species exhibit different abilities to accumulate metals). Only plant species that showed a positive metal uptake resulting from sludge applications can be considered for evaluation of factors that affect response. Environmental factors that influence the metal accumulations are reviewed in detail in Chapters 2 and 3. Where possible, these differences have been recorded along with data presented.

The recognition of these constraints served as the impetus for the USDA, CSRS Regional Technical Committee, W-124, to conduct a "uniform" field study at 15 locations across the United States in 1979 (see Chapter 2 for details). Data from this study were used in our delineation.

CUMULATIVE EFFECTS FROM ANNUAL SLUDGE APPLICATIONS

From the data provided by several researchers (Vlamis, et al., 1985; Soon and Bates, 1981; Chang et al., 1983; Hinesly et al., 1984) it was apparent that cumulative effects from annual sludge applications may be broken down into two categories according to the metal inputs, namely: (1) zinc and cadmium when introduced with sludge at high levels (>100 kg Zn/ha/yr and >1 kg Cd/ha/yr) resulted in increases of plant tissue metal over the years of sludge application, but the rate of increase decreased with time; and (2) typical sludges applied at agronomic rates to satisfy N requirement for crop growth cause Cd and Zn concentrations in plants to become greater than those of the control but Cd and Zn contents of plant tissue remained at a low, nearly constant level with each successive sludge application.

When sludges were applied at rates equivalent to 8.6 kg Cd/ha and 714 kg Zn/ha annually over a 7-year period, the concentration of Cd in the straw of barley increased from 0.26 at the first year to 3.39 mg/kg in the seventh growing season (Table 18, Vlamis et al., 1985). At the the same time the Zn concentration of the plant tissue increased from 113 to 820 mg/kg. During the course of the experiment, pH of the sludge-treated soil gradually decreased from 5.5 to 4.8 which could account for some of the

Table 18. Cd and Zn contents of plant tissues when sludges were applied annually at high rates.

Metal/Crop	Plant parts	Total metal inputs (kg/ha)	pH Initial/final	No. of successive annual applications — Metal in plant tissue (mg/kg)										Reference
				1	2	3	4	5	6	7	8	9	10	
Cd/Barley	straw	60	5.5/4.8	0.26	0.23	0.55	0.63	0.85	1.61	3.39	-	-	-	
Barley	straw	0	5.5/5.5	0.09	0.04	0.07	0.06	0.04	0.10	0.06	-	-	-	Vlamis et al.1985
Corn	stover	5.44	7.5/6.7	0.38	0.54	0.80	0.70	0.53	0.36	0.39	0.57	-	-	Soon & Bates 1981
Brom-grass	above ground	6.08	7.4/7.2	0.09	0.39	0.23	0.20	0.28	0.27	0.30	0.45	-	-	Soon & Bates 1981
Swiss chard	above ground	80	7.0/6.2	0.9	2.7	3.2	4.2	6.9	7.1	9.4	4.4	13.1	18.0	Chang & Page 1985
Swiss chard	above ground	20	7.1/6.5	0.5	0.7	1.2	1.0	2.0	3.3	5.5	3.2	7.2	8.7	Chang & Page 1985
Swiss chard	above ground	0	7.2/7.5	0.3	0.3	0.2	0.1	0.4	0.9	1.2	0.6	0.8	1.7	Chang & Page 1985
Zn/Barley	straw	5000	5.5/4.8	113	150	248	341	402	455	820	-	-	-	Vlamis et al.1985
Barley	straw	0	5.5/5.1	72	58	71	51	37	116	68	-	-	-	
Corn	stover	680	7.5/6.8	41	69	103	88	85	77	74	65	-	-	Soon & Bates 1981
Brom-grass	above ground	672	7.4/6.9	28	41	39	34	40	42	44	65	-	-	Soon & Bates 1981
Swiss chard	above ground	6400	7.0/6.2	105	191	216	249	324	209	275	372	844	997	Chang & Page 1985
Swiss chard	above ground	1600	7.1/6.5	79	90	111	116	110	155	146	261	345	567	Chang & Page 1985
Swiss chard	above ground	0	7.2/7.5	67	50	40	39	66	67	56	52	68	34	Chang & Page 1985

increase in metal accumulations by the barley. While the Cd and Zn accumulation in the vegetative part of barley was substantial, the barley grain harvested from sludge-treated soils contained considerably lower levels of Cd and Zn and frequently were not significantly different than those of the control. Swiss chard shows a similar Cd and Zn uptake response, except in greater amounts. With an average input of 8 kg Cd/ha annually, Swiss chard took up 0.9 mg Cd/kg plant tissue the first year and its Cd content increased to 18.0 mg/kg by the tenth year. Again, as with the barley/sludge system, the soil pH decreased from its original 7.0 to 6.5. The long-term effects of sludge applications on the Cd and Zn levels in affected soils and Swiss chard are illustrated in Fig. 6 and 7 (Chang and Page, 1985, Personal communication).

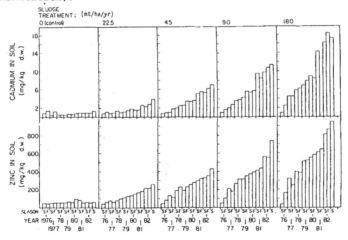

Fig. 6. Cd and Zn concentrations of composted sludge treated Ramona sandy loam (Chang and Page, 1985).

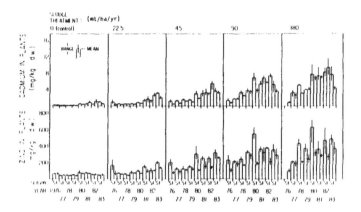

Fig. 7. Cd and Zn content of Swiss chard harvested from soils receiving biannual (spring and fall) · sludge application from 1976-1983 (Chang and Page, 1985).

Soon and Bates (1981) measured Cd and Zn contents of corn and bromegrass on sludge-treated plots with a total accumulative Cd and Zn additions of 5.6 and 680 kg/ha, respectively, over an 8-year period. As with barley and Swiss chard, addition of sludge resulted in an increased metal content in the corn and bromegrass. Although the Cd and Zn inputs were high, the successive additions of sludge did not result in a continuous increase of Cd and Zn concentrations in plants. The use of iron (Fe^{+++}) and aluminum (Al^{+++}) treated sludges in this experiment might have affected the results. Table 19 summarizes the Cd and Zn contents in crops that were grown with sludges applied to satisfy nitrogen requirements of plant growth. In all cases, the concentration of Cd and Zn in the affected plant tissue remained

Table 19. Cd and Zn contents of plant tissue when sludges were applied at agronomic rates.

Metal/Crop	Plant parts	Yrs. of sludge applica- tion	Total metal inputs	pH Initial/ final	No. of successive annual applications								Reference
					1	2	3	4	5	6	7	8	
					Metal in plant tissue (mg/kg)								
Cd/Barley	straw	7	2	5.5/7.0	0.23	0.07	0.16	0.20	0.12	0.26	0.24	-	Vlamis et al. 1985
Barley	straw	7	0	5.5/5.8	0.08	0.04	0.06	0.08	0.09	0.08	0.04	-	
Corn	stover whole	8	0.72	7.4/7.3	--	0.30	0.29	0.27	0.25	0.17	0.16	0.18	Soon & Bates 1981
Brome- grass	plant	8	1.6	7.4/7.4	0.04	0.16	0.08	0.12	0.11	0.08	0.12	0.11	Soon & Bates 1981
Barley	grain	6	5.5	6.1/6.7	0.07	0.05	0.04	0.04	0.04	0.05	-	-	Chang et al. 1983
Barley	grain	6	0	6.3/7.1	0.07	0.04	0.03	0.04	0.04	0.04	-	-	
Barley	grain	6	5.5	7.1/7.1	0.06	0.02	0.01	0.04	0.04	0.05	-	-	Chang et al. 1983
Barley	grain	6	0	7.1/7.1	0.03	0.02	0.05	0.04	0.04	0.04	-	-	
Zn/Barley	straw	7	131	5.5/7.0	66	46	52	46	44	93	54	-	Vlamis et al. 1985
Barley	straw	7	0	5.5/5.8	45	41	52	36	30	98	57	-	
Corn	stover whole	8	112	7.5/7.3	26	23	35	30	31	30	29	27	Soon & Bates 1981
Brome- grass	plant	8	192	7.4/7.4	20	24	25	21	27	24	25	12	Soon & Bates 1981
Barley	leaf	6	80	6.1/6.7	19	22	23	32	32	47	-	-	Chang et al. 1983
Barley	leaf	6	0	6.3/7.0	16	13	25	21	18	22	-	-	
Barley	leaf	6	80	7.1/6.9	24	20	17	29	25	26	-	-	Chang et al. 1983
Barley	leaf	6	0	7.1/7.1	20	14	18	22	21	21	-	-	

constant over the years of application. The levels of Cd and Zn were greater for plants in the sludge-treated plots than the control plots.

There appeared to be a slight but statistically significant increase of Cu and Ni in plants grown on sludge-treated soils when they were compared to plants grown on untreated soils. Their levels did not appear to rise annually with the successive sludge inputs (Soon and Bates, 1981, Table 20). The crops (corn and bromegrass) were grown on calcareous soils which undoubtedly reduced plant availability of the added metals. An earlier report (Vlamis et al., 1978) on sludge application to a non-calcareous soil, however, supported the observations that Cu and Ni are generally not accumulated in plant tissue. In this study 324 kg Cu/ha and 97 kg Ni/ha were applied in the form of sludges over a three-year period and the barley grown on sludge-treated soils did not accumulate significant amounts of Cu and Ni. Studies by other investigators (Chaney, 1985, Personal communication) also showed little detrimental effect to plants at Cu and Ni input levels considerably higher than those reported by Soon and Bates (1981) and Vlamis et al. (1978). In sludge treated soils maintained at pH > 6.0, phytotoxicity from sludge applied Cu and Ni accumulation has rarely been reported (Marks et al., 1980).

From the data reviewed, it is apparent that potentially harmful metal elements can accumulate in plant tissue through land applications of sludge. Amounts absorbed by plants, however, are small and usually accounted for <1% of the inputs from sludge.

SINGLE VS. MULTIPLE APPLICATION

Depending on the way sludges are applied, plants often respond to Cd and Zn introduced into soils in a different manner. Response curves (i.e., metal input from sludge applications vs. metal levels in plants) generated from single sludge additions usually have steeper slopes than response curves generated from multiple sludge additions which have the same total input spanned over a period of time. This would imply that the relationship between total applied metal and the resulting metal content in

Table 20. Copper and Ni contents of plant tissues from sludge-treated soils.

Metal	Crop	Plant parts	Yrs. of sludge application	Total metal inputs	pH Initial/final	No. of successive annual applications								Reference
						1	2	3	4	5	6	7	8	
						Metal in plant tissue (mg/kg)								
Ni	Corn	stover	8	507	7.5/8.0	-	1.5	1.5	0.5	2.3	1.0	1.2	1.9	Soon & Bates 1981
	Corn	stover	8	63	7.5/7.8	-	1.0	1.6	0.4	1.9	0.7	0.6	0.7	
	Brom-grass	whole	8	624	7.4/8.0	-	0.2	1.2	0.6	6.0	4.3	4.2	4.8	Soon & Bates 1981
	Brom-grass plant	whole whole	8	156	7.4/7.7	-	0.3	1.6	0.5	3.5	2.0	2.2	1.9	
Cu	Corn	stover	8	354	7.5/6.7	-	12	10	7	11	8	9	8	Soon & Bates 1981
	Corn	stover	8	88	7.5/7.3	-	8	8	6	9	6	6	6	
	Brom-grass	whole	8	392	7.5/6.7	-	11	15	14	10	15	14	14	Soon & Bates 1981
	Brom-grass plant	whole plant	8	98	7.4/7.3	-	8	10	9	10	10	7	8	

plants is not necessarily unique. Based on the results of a greenhouse pot experiment, the relationship appeared to be a function of the annual application rate (Ryan, 1985, Personal communication).

Results from the W-124 experiment were used to illustrate the patterns of metal concentrations in plants with multiple sludge applications and a single sludge addition which had the equal total input. To summarize the data from various locations into a single diagram, we converted metal concentrations in plant tissue into "relative metal increment of plant tissue" (RMI) which is the ratio of metal increment of plants for a given year (i.e., metal concentration of affected plant tissue minus metal concentration of the control plant) to the first year metal increment of plants receiving 20 mt/ha treatment.

The data may be presented graphically. Under the multiple sludge applications, the line of RMI=0 represents the metal concentration of plants equal to the background metal concentration (Fig. 8). The line RMI=1 represents non-additive effect which indicate, with subsequent sludge inputs, the increment of metal concentration in affected plant tissue are equivalent to that of the first year. The additive effect of multiple sludge inputs on metal contents of plant tissue is represented by the 1 to 1 line that passes through the origin. There was a wide range in the relative metal increments of each location, and in one occasion the RMI even exceeded the strictly additive regime. The mean annual RMI for all locations, however, were approximately 1 (0.86-1.08) indicating non-additive effects due to multiple sludge applications. Sometimes, the relative metal increments of the plant tissue in subsequent years was significantly lower than increments of the first year.

A large single sludge application (100 mt/ha in this case) produced a high plant tissue metal concentration in the crop

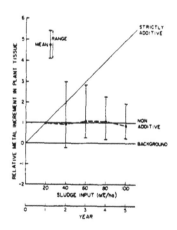

Fig. 8. Relative Zn increments of barley leaf receiving annual sludge addition of 20 mt/ha for five years (calculated with data from 11 of the 15 experimental sites of W-124).

immediately following the sludge application (Fig. 9). This
large single sludge application produced a sharp rise of Zn
levels in plant tissue. In 3 out of the 11 cases, the first year
metal increment of plant tissue exceeded those calculated by the
strictly additive rule with first year metal increment of the 20
mt/ha/yr as the reference point. The RMI of successive crops
from the single sludge addition, however, decreased. By the time
when inputs from the multiple applications had reached the same
amounts as with the single application (year 5 in this case), the
plant tissue metal increment of the single sludge application was
not significantly different from that of the multiple sludge
applications.

METAL ACCUMULATIONS FOLLOWING TERMINATION OF SLUDGE APPLICATIONS

In the early days of land application studies, several
researchers hypothesized that organically complexed metals in
soils were less available to plants than uncomplexed metals.
When sludge applications were terminated, soil microbial activity
would reduce organic matter levels of the sludge-amended soils
resulting in a higher availability of sludge-borne metals
(Chaney, 1973; Haghiri, 1974; Brown, 1975). But the long-term
observations made in field experiments show that the plant
availability of metals in sludge-treated soils either remained
unchanged or was reduced with time after cessation of sludge
applications (Touchton et al., 1976; Dowdy et al., 1978).
Data from a field experiment in Illinois, where sludge was
applied annually for three consecutive years at the agronomic
nitrogen rates on silt loam soils, showed that Cd and Zn concentrations in leaves, stover and grain of corn were increased
significantly by the sludge additions. After sludge applications
were terminated, concentrations of these metals in aerial parts

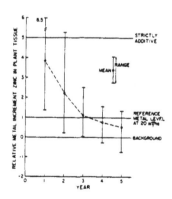

Fig. 9. Relative Zn increments of barley leaf receiving 100 mt/ha one-
time sludge appliction (calculated with data from 11 of the 15
experimental sites of W-124).

of corn plants, although still higher than the control, decreased with each successive corn crop (Hinesly et al., 1984b). Three years after sludge additions ceased, Cd concentrations of corn grain from the sludge-affected soils had receded to levels similar to those from control plots and levels of the metals in leaves and stover from sludge-treated plots were slightly higher than those from control plots. King and Dunlop (1982), Webber and Beauchamp (1979) and Dowdy et al. (1978) reported similar patterns of metal accumulation in plant tissue.

Crops grown on a soil which received annual sludge applications exhibited a slight but significant increase in Cd and Zn concentrations of plant tissues (Hyde et al., 1979). Immediately following the termination of sludge applications at this location, Chang et al. (1982) grew two winter wheat crops and observed that Cd and Zn concentrations of wheat grain and straw from sludge-treated plots were slightly higher than those from control plots. Concentrations of these metals. however, were well within normal ranges of concentrations found in wheat grown on uncontaminated soils. Similar results were found by Hinesly (1985, Personal communication) for Cd, Cu, Ni, and Zn contents in soybeans and wheat grown on plots of silt loam soil where sludge applications were terminated after four and six years of annual applications.

Even for soils that received repeated heavy sludge applications and for plants that were known to accumulate metals, there was little indication that the availability of sludge-borne heavy metals would rise upon termination of sludge applications. In one field trial, the spring and fall split applications of composted sludge at rates ranging from 22.5 to 180 mt/ha/yr on one-half of the experimental field was discontinued after the 6th year. The Cd and Zn contents of plants (Swiss chard and radish) harvested from the area no longer receiving sludges remained elevated but concentrations were lower than those obtained from the area where sludge applications continued (Chang and Page, 1985, Personal communication). For the six croppings (3 years) following the interruption of sludge applications, the metal concentrations of harvested plants remained at levels similar to or less than those at the time sludge application was terminated (Tables 21 to 26).

Based on the data, there is no evidence that the bioavailability of metals in sludge-treated soils will rise with time after terminating the sludge application. Unless chemical conditions of the sludge-treated soils are altered or a metal sensitive plant species is planted, there is no indication that plant uptake of metals should increase with time following termination of sludge application.

Table 21. Cadmium concentrations (mg/kg d.w.) of Swiss chard grown on sludge-treated soils
(Chang and Page, 1985, Personal communication).

Year	Season	Control	22.5 mt/ha/yr* Continued Application	22.5 mt/ha/yr* Terminated	45 mt/ha/yr* Continued Application	45 mt/ha/yr* Terminated	90 mt/ha/yr* Continued Application	90 mt/ha/yr* Terminated	180 mt/ha/yr* Continued Application	180 mt/ha/yr* Terminated
1976	Spring	0.72	1.05	-	1.28	-	1.30	-	-	-
	Fall	0.28	0.35	-	0.52	-	0.88	-	0.88	-
1977	Spring	0.20	0.58	-	1.35	-	1.75	-	3.28	-
	Fall	0.25	0.40	-	0.72	-	1.78	-	2.72	-
1978	Spring	0.20	0.48	-	1.60	-	3.38	-	5.12	-
	Fall	0.20	0.50	-	1.20	-	2.58	-	3.18	-
1979	Spring	0.28	0.62	-	1.55	-	3.75	-	4.02	-
	Fall	0.12	0.52	-	0.98	-	3.48	-	4.15	-
1980	Spring	0.52	0.90	-	3.05	-	6.85	-	7.75	-
	Fall	0.42	0.72	-	1.95	-	3.80	-	6.88	-
1981	Spring	0.50	1.45	-	3.15	-	5.40	-	7.35	-
	Fall	0.90	1.75	-	3.30	-	6.82	-	7.12	-
1982	Spring	0.50	1.35	1.70	3.15	3.80	5.75	6.10	8.32	11.10
	Fall	1.20	2.68	2.30	5.50	5.20	7.45	6.70	9.38	8.70
1983	Spring	0.70	3.18	2.00	3.72	3.80	5.29	4.60	7.65	7.20
	Fall	0.50	2.02	1.30	3.32	2.50	3.65	3.80	4.38	3.70
1984	Spring	1.40	4.82	2.30	5.68	5.60	6.30	6.80	9.32	5.80
	Fall	0.8	3.00	1.40	7.20	4.80	9.80	7.40	13.10	10.70
1985	Spring	1.4	3.30	1.00	11.3	3.60	18.30	10.90	21.60	17.20
	Fall	1.5	4.70	2.80	8.70	5.60	13.60	8.60	18.00	9.50

*Experimental field was split after Fall 1981 where one-half continued to receive sludge and the other half no longer received sludge but was cropped.

Table 22. Zinc concentrations (mg/kg d.w.) of Swiss chard grown on sludge-treated soils
(Chang and Page, 1985, Personal communication).

Year	Season	Control	22.5 mt/ha/yr* Continued Application	22.5 mt/ha/yr* Terminated	45 mt/ha/yr* Continued Application	45 mt/ha/yr* Terminated	90 mt/ha/yr* Continued Application	90 mt/ha/yr* Terminated	180 mt/ha/yr* Continued Application	180 mt/ha/yr* Terminated
1976	Spring	65	167	-	170	-	229	-	-	-
	Fall	67	67	-	79	-	105	-	105	-
1977	Spring	45	72	-	124	-	215	-	215	-
	Fall	50	60	-	90	-	189	-	191	-
1978	Spring	48	62	-	127	-	241	-	432	-
	Fall	40	47	-	111	-	172	-	216	-
1979	Spring	51	85	-	149	-	332	-	378	-
	Fall	39	78	-	116	-	289	-	249	-
1980	Spring	81	101	-	301	-	550	-	633	-
	Fall	66	71	-	110	-	192	-	324	-
1981	Spring	76	152	-	297	-	368	-	366	-
	Fall	67	96	-	155	-	322	-	290	-
1982	Spring	61	107	127	251	278	382	337	475	465
	Fall	67	106	88	146	174	213	240	275	282
1983	Spring	52	203	103	320	268	361	313	490	404
	Fall	54	138	94	261	225	293	386	372	304
1984	Spring	84	202	95	176	218	319	348	373	230
	Fall	66	188	98	345	290	554	554	844	730
1985	Spring	73	232	101	389	252	725	385	1000	548
	Fall	37	209	99	567	218	768	510	997	562

*Experimental field was split after Fall 1981 where one-half continued to receive sludge and the other half no longer received sludge but was cropped.

Table 23. Cadmium concentrations (mg/kg d.w.) of radish leaf grown on sludge-treated soils
(Chang and Page, 1985, Personal communication)

Year	Season	Control	22.5 mt/ha/yr* Continued Application	Termi- nated	45 mt/ha/yr* Continued Application	Termi- nated	90 mt/ha/yr* Continued Application	Termi- nated	180 mt/ha/yr* Continued Application	Termi- nated
1976	Spring	0.2	0.4	-	0.4	-	0.5	-	-	-
	Fall	0.2	0.6	-	0.9	-	1.1	-	1.5	-
1977	Spring	0.3	0.6	-	0.9	-	1.5	-	3.4	-
	Fall	0.5	0.7	-	1.9	-	2.8	-	5.0	-
1978	Spring	0.2	0.2	-	1.2	-	1.4	-	1.9	-
	Fall	0.3	1.0	-	1.4	-	2.0	-	3.2	-
1979	Spring	0.5	2.0	-	2.5	-	3.1	-	3.9	-
	Fall	0.7	2.5	-	3.6	-	5.9	-	7.4	-
1980	Spring	1.0	1.4	-	2.0	-	3.2	-	5.2	-
	Fall	0.6	1.8	-	3.8	-	3.8	-	8.8	-
1981	Spring	1.1	1.8	-	2.7	-	5.4	-	7.0	-
	Fall	1.7	2.3	-	4.2	-	6.1	-	8.4	-
1982	Spring	0.8	2.1	1.2	3.1	1.9	5.9	2.9	6.6	6.3
	Fall	1.8	3.7	3.1	5.6	3.5	7.2	4.8	6.2	3.0
1983	Spring	1.4	2.8	2.1	3.7	2.7	4.9	3.6	6.8	6.1
	Fall	0.8	2.2	1.5	2.9	2.2	3.7	2.4	5.3	3.7
1984	Spring	0.9	2.5	1.2	6.6	2.6	8.0	3.2	8.8	7.1
	Fall	0.6	2.8	1.1	6.7	2.1	8.5	3.6	14.2	7.5
1985	Spring	0.7	4.2	1.7	11.8	3.7	13.7	9.5	14.9	11.3
	Fall	0.6	3.4	1.6	5.5	3.2	7.9	4.7	10.8	7.4

*Experimental field was split after Fall 1981 where one-half continued to receive sludge and the other half no longer received sludge but was cropped.

Table 24. Cadmium concentrations (mg/kg d.w.) of radish tuber grown on sludge-treated soils
(Chang and Page, 1985, Personal communication).

Year	Season	Control	22.5 mt/ha/yr* Continued Application	Termi- nated	45 mt/ha/yr* Continued Application	Termi- nated	90 mt/ha/yr* Continued Application	Termi- nated	180 mt/ha/yr* Continued Application	Termi- nated
1976	Spring	0.2	0.2	-	0.2	-	0.2	-	-	-
	Fall	0.1	0.2	-	0.3	-	0.4	-	0.7	-
1977	Spring	0.1	0.3	-	0.3	-	0.4	-	0.9	-
	Fall	0.2	0.4	-	0.7	-	0.9	-	1.9	-
1978	Spring	0.2	0.3	-	0.4	-	0.5	-	0.7	-
	Fall	0.1	0.5	-	0.7	-	1.2	-	1.8	-
1979	Spring	0.2	0.4	-	0.5	-	0.7	-	1.1	-
	Fall	0.3	1.0	-	1.1	-	1.5	-	2.2	-
1980	Spring	0.3	0.5	-	0.5	-	0.8	-	1.0	-
	Fall	0.3	0.8	-	1.4	-	2.5	-	2.7	-
1981	Spring	0.3	0.4	-	0.6	-	0.9	-	1.0	-
	Fall	0.5	0.5	-	0.6	-	0.8	-	0.9	-
1982	Spring	0.4	0.6	0.4	1.0	0.7	1.4	1.0	1.4	1.4
	Fall	0.4	0.5	0.4	0.8	0.5	1.3	0.9	1.4	1.2
1983	Spring	0.8	0.7	0.7	0.8	0.7	0.7	0.7	0.9	0.7
	Fall	0.6	1.1	0.9	1.6	1.1	2.1	1.5	2.2	2.1
1984	Spring	0.3	0.9	1.0	1.0	0.8	1.2	0.9	1.2	1.0
	Fall	0.3	1.1	0.6	1.6	1.0	1.9	1.2	2.7	2.3
1985	Spring	0.6	1.1	1.0	1.0	1.1	2.5	1.3	3.5	2.6
	Fall	0.3	1.9	1.1	3.1	1.9	4.0	3.8	5.6	4.3

*Experimental field was split after Fall 1981 where one-half continued to receive sludge and the other half no longer received sludge but was cropped.

Table 25. Zinc concentrations (mg/kg d.w.) of radish leaf grown on sludge-treated soils (Chang and Page, 1985, Personal communication).

Year	Season	Control	22.5 mt/ha/yr*		45 mt/ha/yr*		90 mt/ha/yr*		180 mt/ha/yr*	
			Continued Application	Termi-nated	Continued Application	Termi-nated	Continued Application	Termi-nated	Continued Application	Termi-nated
1976	Spring	55	56	--	62	-	55	--	-	--
	Fall	42	58	-	75	-	78	-	173	--
1977	Spring	45	46	--	56	-	73	--	79	-
	Fall	55	82	-	64	-	102	-	138	--
1978	Spring	46	46	-	71	--	93	--	121	-
	Fall	39	64	-	90	-	134	-	177	--
1979	Spring	38	52	-	74	--	129	-	198	-
	Fall	4.7	95	-	232	-	223	-	254	-
1980	Spring	42	58	-	81	--	129	-	193	-
	Fall	40	66	-	121	-	201	-	294	-
1981	Spring	46	73	-	117	--	216	-	266	--
	Fall	50	84	-	134	-	176	-	275	-
1982	Spring	45	74	56	109	78	206	120	234	200
	Fall	42	79	63	142	82	192	116	283	184
1983	Spring	41	74	51	107	76	167	110	275	196
	Fall	48	95	57	159	90	230	133	354	225
1984	Spring	50	97	62	195	97	316	180	421	244
	Fall	70	147	73	262	120	383	171	627	286
1985	Spring	32	103	53	186	85	305	143	452	231
	Fall	44	106	62	166	93	232	120	353	179

*Experimental field was split after Fall 1981 where one-half continued to receive sludge and the other half no longer received sludge but was cropped.

Table 26. Zinc concentrations (mg/kg d.w.) of radish tuber on sludge-treated soils (Chang and Page, 1985, Personal communication).

Year	Season	Control	22.5 mt/ha/yr*		45 mt/ha/yr*		90 mt/ha/yr*		180 mt/ha/yr*	
			Continued Application	Termi-nated	Continued Application	Termi-nated	Continued Application	Termi-nated	Continued Application	Termi-nated
1976	Spring	34	38	-	33	-	39	-	-	-
	Fall	48	57	-	70	-	47	-	92	-
1977	Spring	38	31	-	37	-	51	-	61	-
	Fall	31	46	-	56	-	62	-	84	-
1978	Spring	20	32	-	57	-	94	-	54	-
	Fall	42	63	-	86	-	124	-	158	-
1979	Spring	36	59	-	56	-	71	-	88	-
	Fall	37	46	-	71	-	97	-	135	-
1980	Spring	34	41	-	51	-	63	-	76	-
	Fall	49	58	-	82	-	131	-	144	-
1981	Spring	33	39	-	52	-	70	-	77	-
	Fall	37	40	-	52	-	67	-	79	-
1982	Spring	42	50	46	71	57	99	73	105	93
	Fall	31	40	30	53	38	80	46	95	72
1983	Spring	30	42	36	44	40	60	41	82	58
	Fall	46	58	39	75	51	101	76	117	101
1984	Spring	31	44	40	69	48	89	63	113	76
	Fall	40	84	50	118	68	154	90	224	157
1985	Spring	24	38	33	68	39	94	49	116	72
	Fall	42	78	52	88	64	110	79	132	84

*Experimental field was split after Fall 1981 where one-half continued to receive sludge and the other half no longer received sludge but was cropped.

CONCLUSIONS

1. Application of Cd and Zn to soils from municipal sludge will cause the Cd and Zn concentration of plants grown on these soils to exceed those of the untreated controls. When the sludge is applied at rates to satisfy the N requirement of the crop grown the Cd and Zn contents of plant tissue remain at nearly constant levels with successive sludge applications.

2. In sludge treated soils maintained at pH >6.0, Cu and Ni contents of the tissue from plants grown on these soils may become slightly elevated. Phytotoxicity from sludge-applied Cu and Ni, however, has rarely been reported.

3. Given adequate time for sludge to equilibrate with the soil, metal concentration of the affected plant tissue would be determined by the total amounts of metals in the soil and would not be affected by the methods of sludge application (e.g., single addition vs. multiple applications to yield the same total application as the single addition).

4. Plant availability of sludge-borne metals is highest during the first year sludge is applied. Using the first year response curve generated by a large single sludge addition will overestimate metal accumulation in vegetative tissue from plants grown in well stabilized sludge/soil systems.

5. There are no field data to indicate that trace element concentration in plant tissue will rise after the termination of sludge applications if chemical conditions of the soil remain constant. Cadmium and zinc levels of plants grown in soils which are no longer receiving sludges either were not significantly different from the pretreatment levels or decreased with time.

REFERENCES

Brown, R. E. 1975. Significance of trace metals and nitrates in sludge soils. J. Water Pollut. Control Fed. 47:2863-2875.

Chaney, R. L. 1973. Crop and food chain effects of toxic elements in sludges and effluents. p. 129-141. In Proc. of the joint conf. on recycling municipal sludges and effluents on land, Champaign, IL. 9-13 July 1973. National Assoc. State Univ. and Land Grant Colleges. Washington, D.C.

Chaney, R. L. 1985. Personal communication, USDA-ARS-NER, Beltsville, MD 20705.

Chang, A. C., A. L. Page, and F. T. Bingham. 1982. Heavy metal absorption by winter wheat following termination of cropland sludge applications. J. Environ. Qual. 11:705-708.

Chang, A. C., A. L. Page, J. E. Warneke, M. R. Resketo, and T. E. Jones. 1983. Accumulation of cadmium and zinc in barley grown on sludge-treated soils: A long-term field study. J. Environ. Qual. 12:391-397.

Chang, A. C. and A. L. Page. 1985. Personal communication. Dept. of Soil and Environmental Sciences, Univ. of California, Riverside, CA 92521.

Dowdy, R. H., W. E. Larson, J. M. Titrud, and J. J. Latterell. 1978. Growth and metal uptake of snap beans grown on sewage sludge-amended soils, a four year field study. J. Environ. Qual. 7: 252-257.

Haghiri, F. 1974. Plant uptake of cadmium as influenced by cation exchange capacity, organic matter, zinc, and soil temperature. J. Environ. Qual. 3:180-183.

Hinesly, T. D., E. L. Ziegler, and G. L. Barrett. 1979. Residual effects of irrigating corn with digested sewage sludge. J. Environ. Qual. 8:35-38.

Hinesly, T. D., K. E. Redborg, R. I. Pietz and E. L. Ziegler. 1984. Cadmium and zinc uptake by corn (Zea mays L.) with repeated application of sewage sludge. J. Agric. Food Chem. 32:155-163.

Hinesly, T. D. 1985. Personal communication, Dept. of Agronomy, Univ. of Illinois, Urbana, IL 61801.

Hyde, H. C., A. L. Page, F. T. Bingham and R. J. Mahler. 1979. Effect of heavy metals in sludge on agricultural crops. J. Water Pollut. Control Fed. 51:2475-2486.

King, L. D., and W. R. Dunlop. 1982. Application of sewage sludge to soils high in organic matter. J. Environ. Qual. 11: 608-616.

Logan, T. J. and R. L. Chaney. 1983. Utilization of municipal wastewater and sludge on land - Metals. pp. 235-326. In A. L. Page et al. (ed.). Proc. of the workshop on utiliza- tion of municipal wastewater and sludge on land. Univ. of California, Riverside, CA.

Logan, T. J., A. C. Chang, A. L. Page, and T. J. Ganje. 1987. Accumulation of selenium in crops grown on sludge-treated soil. J. Environ. Qual. (in press).

Marks, M. J., J. H. Williams and C. G. Chumbley. 1980. Field experiments testing the effects of metal contaminated sludges on some vegetable crops. p. 235-251. In Inorganic Pollu- tion and Agriculture. Min. Agr. Fish. Food Reference Book 326. HMSO, London.

Ryan, J. A. 1985. Personal communication, Municipal Environmental Research Laboratory, U.S. EPA, Cincinnati, OH 45268.

Soon, Y. K. and T. E. Bates. 1981. Land disposal of sewage sludge, a summary of research results, 1972-1980. Dept. of Land Resources Sci., Univ. of Guelph, Ontario, Canada.

Soon, Y. K. and T. E. Bates. 1985. Molybdenum, cobalt and boron uptake from sewage-sludge-amended soils. Can. J. Soil Sci. 65:507-517.

Touchton, J. T., L. D. King, H. Bell and H. D. Morris. 1976. Residual effect of liquid sludge on coastal bermudagrass and soil chemical properties. J. Environ. Qual. 5:161-164.

Vlamis, J., D. E. Williams, J. E. Corey, A. L. Page, and T. J. Ganje. 1985. Zinc and cadmium uptake by barley in field plots fertilized seven years with urban and suburban sludge. Soil Sci. 139:81-87.

Vlamis, J., D. E. Williams, K. Fong, and J. E. Corey. 1978. Metal uptake by barley from field plots fertilized with sludge. Soil Sci. 126:49-55.

Webber, L. R. and E. G. Beauchamp. 1979. Cadmium concentrations and distribution in corn (Zea mays L.) grown on a calcareous soil for three years after three annual sludge applications. J. Environ. Sci. Health B14(5):454-474.

CHAPTER 5

TRANSFER OF SLUDGE-APPLIED TRACE
ELEMENTS TO THE FOOD CHAIN

Rufus L. Chaney
USDA-ARS-NER, Beltsville, Maryland

Randall J. F. Bruins
U.S. Environmental Protection Agency, Cincinnati, Ohio

Dale E. Baker
Pennsylvania State University, University Park, Pennsylvania

Ronald F. Korcak
USDA, Beltsville, Maryland

James E. Smith
U.S. Environmental Protection Agency, Cincinnati, Ohio

Dale Cole
University of Washington, Seattle, Washington

INTRODUCTION

Assessment of the likelihood of risks to humans, livestock, and wildlife from potentially toxic constituents in sewage sludge applied to land requires a knowledge of the potential for transfer of each constiuent from the sludge or sludge-soil mixture to crops and to animals (including humans) which ingest sludge, sludge-soil mixture, or crops grown on the sludge-amended soil. Transfer of sludge constituents from soil to crops is predominantly a function of: (1) the constituent; (2) soil pH; (3) characteristics of the applied sludge and cumulative sludge application rate; and (4) the crop species and cultivar grown. Each of these factors can be associated with a 2-fold or greater change in plant concentration of some trace elements.
    Earlier attempts to estimate food-chain transfer have used plant uptake slopes obtained by linear regression of the consti- tuent's concentration in edible crop tissue and the amount of the

constituent applied [(mg constituent/kg dry crop) per (kg constituent applied/ha)]. In the long-term, plant uptake of sludge-applied Cd and Zn is a curvilinear (plateauing) response to the cumulative application rate of applied constituent in a sludge. Further, the plateau reached is a function of the concentration of the constituent in sludge and other sludge properties such as Fe added during sludge processing.

Estimation of food-chain transfer is critical to valid estimation of the potential for risk. At present, these estimates are best made by considering (1) the relative increased uptake of constituents by various crops from sludge-amended soil under responsive conditions; (2) the rate of ingestion of different crops by the U.S. population (e.g., Pennington, 1983); (3) the demonstrated bioavailability of the increased amounts of an element in sludge-grown crops or ingested sludge; and (4) an appropriate transfer coefficient from sludge-amended soil to edible crop tissues [i.e., the increase in crop trace element residue (above that in the same crop grown on background soils) at the plateau reached on sludge-amended soil].

Ryan et al. (1982) developed an approach for estimating food-chain transfer of sludge-applied Cd. While this model has to be modified to account for curvilinear response to sludge-applied Cd, many other parts of the approach remain valid. In particular, the aggregate increased food-chain exposure to a sludge-borne constituent for the exposed population can be expressed in terms of a constant times the element transfer coefficient (height of the plateau above that for untreated soil) for an indicator crop such as lettuce. Individuals who grow, on acidic sludge-amended soils, a significant portion of the garden vegetables they ingest are generally believed to be the individuals most likely to have increased dietary Cd due to land application of sludge. For chronic lifetime (50 year) Cd exposure, estimation also relies on U.S. food intake estimates (g fresh weight/day) from Pennington (1983), and solids content of foods from USDA data bases (Adams, 1975). Because response curves or plateaus have not yet been evaluated for every crop consumed by the U.S. population, uptake by various food crops must be represented by the FDA food groups. For Cd, based on many data sources, the response of minor food crops remains well represented by the FDA food groups.

On the other hand, the use of a "Cd-accumulator" crop to represent increased Cd uptake by all crops in a food group has been criticized. For example, lettuce and broccoli were both listed in the leafy vegetable food group, but have at least 10-fold different response slopes. Root vegetables and garden fruits also include high and low element accumulating crops. These wide crop differences have caused an excessive estimated increase in food-chain Cd transfer (providing a hidden safety factor). Thus, food group aggregate transfer response slopes have to be adjusted for the proportion of low and high accumulating crops in each food group. Each food group can still be represented by a constant times the response of lettuce. This approach appears to be appropriate based on the findings of this workshop.

Many studies have shown significantly higher uptake of Cd, Zn, and Ni at lower soil pH compared to neutral soil pH (see Logan and Chaney, 1983; CAST, 1980). Exceptions were reported by Pepper et al. (1983) and Hemphill et al. (1982) in which corn silage was unchanged or slightly higher in Cd on limed sludge-amended soil. One possible reason for this exception is that corn differs from other crops in its mechanism of obtaining Fe from soil. Recent research on the mechanism plants use to obtain Fe from soil has shown that the Graminae (corn, wheat, barley, rice, oat, sorghum, etc.) excrete an organic chelating agent which facilitates Fe diffusion to the root and Fe uptake (Sugiura and Nomoto, 1984; Römheld and Marschner, 1986), while plants in other families do not excrete chelators. These compounds also chelate Cd, Cu, Zn, and other microelements in the presence of Fe (unlike bacterial siderophores, which have very high selectivity for ferric iron). Non-specific chelating agents added to soils are also known to facilitate diffusion and plant uptake of Zn, Cd, and other elements. Thus, Cd uptake by Graminae can have a lesser response to soil pH change than other species, depending on the availability of soil Fe.

Most studies of soil pH effect on metal uptake by crops have studied the pH range from about 5.0 to 7.5. In a report by Francis et al. (1985), S in a coal gasification waste caused soil pH to fall to 4.0, and the ryegrass crop grown on this extremely acidic soil was stunted and very high in metals even though metals in the soil and coal waste were not very high. Extreme acidic soil pH levels associated with severe pH mismanagement may allow crop metal residues not otherwise considered in this report.

Important problems remain in estimating the variance in potential risk due to unusual individual dietary selection patterns, and to individual variation in retention of potentially toxic constituents in foods. For example, it would be useful to know the statistical distribution of increased Cd intake among individuals consuming self-selected diets using crops grown on sludge-amended acid gardens. If these data were available, one could design Cd limits to protect individuals with the 95th percentile of increased exposure. Several papers have noted that the relevant information needed on variation in food intake for evaluation of chronic risk from food Cd is lifetime (50-year) variation in intake of foods, rather than the 1-day variation now available (Dean and Suess, 1985). Although the mean 1-day ingestion of foods for a population is estimated well by a large survey of individual 1-day intakes, the variance in long-term average daily intake is greatly over-estimated by the variance of 1-day intakes for a population (Beaton et al., 1983; Todd et al., 1983; Black, 1982). Beaton et al. (1983) and Sempos et al. (1985) found that intra-individual variance was greater than the inter-individual variance for multiple 1-day observations. Much smaller variances among individuals are associated with long-term dietary intakes, especially if one considers major dietary sub-groups (e.g. vegetarians) separately. Thus, present approaches for estimating the fraction of the population ingesting more Cd than some limit (e.g., 95th percentile, or fraction > 71 µg Cd/day), based on variances in population 1-day food intakes, would greatly overestimate the lifetime Cd intake by individuals.

Further, the protection of individuals varying in retention of ingested elements must consider the effects of important nutritional interactions on element bioavailability. There are several clear examples of this source of error in estimating risk. One is the health effects to Japanese farmers who ingested rice grown in soils rich in Cd and Zn due to contamination by Zn-ore wastes. In contrast to essentially all other food crops, rice is grown in flooded soils. Cadmium and Zn uptake by rice is normally kept very low because insoluble metal sulfides are formed in the soil. However, some metal uptake occurs because the roots obtain oxygen by air channels within the plants. It was found that rice metabolized Cd and Zn differently in these anaerobic rice soils than patterns common to other food crop species in aerobic soils. Although both soil Cd and soil Zn were greatly increased (up to 10 mg Cd/kg and 1200 mg Zn/kg), rice grain Cd rose up to 100-fold while grain Zn was unchanged (Tsuchiya, 1978, page 237). Further, during preparation of polished rice, much of the Ca, Zn, and Fe in brown rice is removed during milling, while a much lower fraction of the Cd is removed (Pedersen and Eggum, 1983; Chino, 1981; Chino and Baba, 1981, Yoshikawa et al., 1977). Lastly, rice Fe has very low bioavailability (Hallberg et al., 1974, 1977). All these factors [crop metal uptake characteristic, food processing, and effect of nutrient (Fe, Zn, and Ca) status on human Cd retention] favored Cd retention by the farmers. Bioavailability of Cd was thus high, and human disease resulted.

In another Cd exposure case, individuals who consumed large amounts of Cd-rich oysters had higher dietary levels of Zn, Fe, and Ca, levels which were more like those in the normal U.S. diet. The oyster Cd had very low apparent bioavailability based on Cd in blood and urine compared to the effect of smoking on Cd in blood and urine (Sharma et al., 1983; McKenzie et al., 1982). These findings suggest that predictions of human retention of Cd from Western-type diets may be less than values currently in use.

Another example comes from the availability of sludge Cu to livestock. In contrast to Cu salts mixed with diets, Cu in sludge fed to livestock has low bioavailability. Generally, ingested sludge lowers liver Cu stores rather than causing Cu toxicity, even though equal levels of soluble Cu salts would poison the animals (Decker et al., 1980; Bertrand et al., 1981; Baxter et al., 1982).

MODELING THE EFFECT OF CROP VARIATION IN INCREASED TRACE ELEMENT ACCUMULATION IN RESPONSE TO SLUDGE APPLICATION

The extent of increase in trace element concentration above control for crops grown on a sludge-amended soil is very strongly affected by crop species. Besides crop species variation in response, sludge Cd concentration and soil pH very strongly

affect the plant Cd:soil Cd relationship. Other factors, such
as crop cultivar and Fe level in the sludge, may have a substantial effect on the plant Cd:soil Cd relationship, while many
other factors have affected results in some studies (soil organic
matter, soil sesquioxides, pH buffering capacity of the soil,
soil fertility, crop mycorrhizal infection, type of N fertilizer,
and climatic factors).

An approach to estimate the relation between sludge-applied
Cd, for example, and increased dietary Cd exposure is to inte-
grate all crop response in terms of Cd uptake by a responsive
reference crop such as lettuce. Relative to the responsive crop,
the increased Cd uptake among crops due to sludge application has
been reasonably consistent. However, few individual experiments
have included a large number of crop species in a responsive
sludge treatment to provide the relative increases in crop levels
of potentially toxic constituents needed for dietary exposure
assessments. The studies by Davis and Carlton-Smith (1980) and
Carlton-Smith and Davis (1983) report the response of many crop
species in one experiment. The conditions of these studies meet
the constraints for appropriate techniques for sludge trace element risk assessment. They grew many crop species (some with
multiple cultivars) on 2 soils collected from long-term sludge
farms in England. They used large pots of soil (10 kg) in the
greenhouse. Soil A had pH 6.7 and contained 5.8% organic matter
and 7.4 mg Cd/kg soil. Soil B had pH 6.8 and contained 26%
organic matter and 68.7 mg Cd/kg soil. Crop Cd ranged from
near zero to about 8 mg/kg dry weight. Relative crop Cd concentrations were similar between these 2 soils, and in good
agreement with other research results.

In an effort to make maximum use of these data, Davis and
Carlton-Smith developed tables of relative element concentrations
for Cd, Zn, Cu, Ni and Pb. The concentration in each crop was
expressed as a percentage of that in the crop with highest uptake
of a given element, and the data were averaged across the 2
soils. The raw data for this study were obtained from Dr. R. D.
Davis. In evaluating the raw data, it was noted that several
data points were outliers, and these data points were deleted.
The geometric mean element concentrations for the 2 soils (Table
27) were calculated; then a normal background Cd concentration
in each crop was subtracted from the geometric mean Cd level in
that crop, and the ratio of Cd in crop X to the mean level of Cd
in lettuce [(Cd in Crop X):(Mean Cd in lettuce)] was calculated
(Table 28). The background Cd levels were estimated for crop
groups, mainly relying on Wolnik et al. (1983a,b; 1985); other
field results were also considered and summarized by Korcak
(1986, Personal communication).

Table 27 shows the geometric mean concentration of Cd, Ni and
Zn in edible crop tissues of vegetable and grain crops. Similar
results for relative uptake for total shoots of forage crops
grown on 2 sludge-amended soils and 1 control soil are presented
for Cd, Cu, Mo, Ni, and Zn in Table 29 from Carlton-Smith and
Davis (1983). (We do not report their Pb data because the crops
were grown near an urban area, and the Pb results are not representative of agricultural production areas).

Table 27.   Trace element concentration in edible plant tissues, and relative
Cd concentrations in edible tissues of crops (dry weight basis);
geometric means from each of 2 long-term sludge-amended soils)
(Davis and Carlton-Smith, 1980).

Crop	Cultivar	Cd	Ni	Zn	Relative Cd Concentration
		—mg/kg dry weight—			
Lettuce	Tom Thumb	8.1	1.2	95.	100.
Lettuce	Webbs	6.9	2.2	96.	85.
Lettuce	Paris White Romaine	6.2	4.0	82.	77.
Spinach	Bloomsdale	5.0	1.4	391.	62.
Kale	Maris Kestrel	1.3	6.0	105.	16.
Cabbage	Greyhound	0.97	5.7	105.	12.
Wheat	Spartacus	0.88	6.4	75.	11.
Mangold	Yellow Globe	0.74	2.2	131.	9.1
Turnip	Bruce	0.74	1.9	45.	9.1
Leek	Musselburgh	0.73	0.91	28.	9.0
Wheat	Sappo	0.62	4.9	75.	7.7
Turnip	Snowball	0.58	2.0	37.	7.2
Rape	Orpal	0.54	8.2	54.	6.7
Onion	White Lisbon	0.52	0.81	40.	6.4
Beetroot	Detroit	0.41	2.3	103.	5.1
Tomato	Moneymaker	0.40	2.0	22.	4.9
Sugarbeet	Sharpes Klein Monobeet	0.35	2.1	130.	4.3
Bettroot	Boltardy	0.34	2.1	76.	4.2
Carrot	Standard Improved	0.33	1.5	38.	4.1
Radish	French Breakfast	0.33	3.1	48.	4.1
Barley	Julia	0.31	4.2	67.	3.8
Parsnip	Giant Exhibition	0.26	3.2	34.	3.2
Barley	Ark Royal	0.25	7.8	69.	3.1
Swede	Acme	0.24	1.2	28.	3.0
Potato	Desiree	0.20	0.66	20.	2.5
Oat	Leander	0.18	7.2	53.	2.2
Squash	Zucchini	0.17	5.8	80.	2.1
Sweet corn	Golden Earley	0.16	0.37	41.	2.0
Sunflower	Tall Single	0.15	11.4	41.	1.9
Maize	Caldera	0.13	1.0	37.	1.6
French bean	Canadian Wonder	0.08	9.9	31.	1.0
Pea	Onward	0.05	4.1	63.	0.6

The relative crop uptake tables remove factors other than crop species and cultivar. From the summarized data, it is not possible to ascertain the effects of other parameters (such as soil organic matter, soil pH, or sludge application rate) on relative metal uptake among crops.

Relative Cd uptake by crops was also evaluated in areas of naturally Cd-rich soils in Salinas Valley, Monterey Co., California (Table 30; Burau, 1980). Many paired samples of mature vegetable produce were obtained along with plow layer soil. The slope of the relationship between crop and soil Cd was reported for crops for which linear regression showed a significant slope. The Cd-enriched soils have quite similar properties, with Cd coming from geologic sources rather than sludge. Table 30 shows slopes and a relative uptake calculation similar to those used in Tables 28 and 29, with leaf lettuce set = 100.

Giordano et al. (1979) reported crop uptake of trace elements as affected by sludge application rate and soil heating. They continued the study for 2 years following application. Their results (unheated soil data) corroborate the extremely low increase of Cd in beans, cabbage, pepper, tomato, and the curcurbit family and low Cd increase in potato.

Table 28. Relative increased Cd concentration in edible tissues of crops
grown on long-term sludge-amended soils. Mean crop Cd from
Table 27 was corrected for normal background levels of Cd
in crops. All increased Cd concentrations were divided by
6.37, the mean corrected Cd concentration in 3 cultivars lettuce
(based on data from Davis and Carlton-Smith, 1980).

Crop	Cultivar	Crop Cd	Back-ground Cd	Increased Crop Cd	Relative Increased Cd Uptake
		——— mg Cd/kg dry weight———			
Lettuce	Tom Thumb	8.1	0.7	7.4	116.
Lettuce	Webbs	6.9	0.7	6.2	97.
Lettuce	Paris White Romaine	6.2	0.7	5.5	86.
Spinach	Bloomsdale	5.0	0.7	3.7	58.
Kale	Maris Kestrel	1.3	0.27	1.0	16.
Wheat	Spartacus	0.88	0.08	0.80	13.
Cabbage	Greyhound	0.97	0.27	0.60	9.4
Wheat	Sappo	0.62	0.08	0.54	8.5
Mangold	Yellow Globe	0.74	0.21	0.53	8.3
Turnip	Bruce	0.74	0.21	0.53	8.3
Leek	Musselburgh	0.73	0.27	0.46	7.2
Rape	Orpal	0.54	0.08	0.46	7.2
Turnip	Snowball	0.58	0.21	0.37	5.8
Onion	White Lisbon	0.52	0.21	0.31	4.9
Barley	Julia	0.31	0.08	0.23	3.6
Beetroot	Detroit	0.41	0.21	0.20	3.1
Barley	Ark Royal	0.25	0.08	0.17	2.7
Sugarbeet	Sharpes Klein Monobeet	0.35	0.21	0.14	2.2
Beetroot	Boltardy	0.34	0.21	0.13	2.0
Radish	French Breakfast	0.33	0.21	0.12	1.9
Oat	Leander	0.18	0.08	0.10	1.6
Carrot	Standard Improved	0.33	0.25	0.08	1.3
Tomato	Moneymaker	0.40	0.32	0.08	1.3
Potato	Desiree	0.20	0.13	0.07	1.1
Sunflower	Tall Single	0.15	0.08	0.07	1.1
Squash	Zucchini	0.17	0.11	0.06	0.9
Parsnip	Giant Exhibition	0.26	0.21	0.05	0.8
Sweet corn	Golden Earley	0.16	0.11	0.05	0.8
Maize	Caldera	0.13	0.08	0.05	0.8
Swede	Acme	0.24	0.21	0.03	0.5
French bean	Canadian Wonder	0.08	0.06	0.02	0.3
Pea	Onward	0.05	0.06	-0.01	-0.2

Background Cd concentrations were based on field grown control crops reported
in many studies, but mainly Wolnik et al. (1983a,b; 1985); Korcak (1986,
Per-sonal communication) summarized these results in a draft report to EPA.

Although it has been clearly shown that crops differ in
uptake of trace elements from the same soil, the biochemical/
physiological basis for crop differences has not been explained.
Basic research by Jarvis et al. (1976) indicated that crops
differed in Cd uptake by roots, and also differed in the fraction
of root Cd translocated to shoots. Recently, Grill et al. (1985)
found that many plant species made a family of cysteine-rich peptides (related to glutathione) which strongly chelate Cd, Zn, Cu,
Pb, and Zn. Their "phytochelatins" may be synthesized in the
fibrous roots and chelate absorbed metals, and thereby protect
root metabolism and reduce trace element translocation to edible
plant tissues (Grill et al., 1985; Rauser and Glover, 1984;
Robinson and Jackson, 1986).

CROP CULTIVAR DIFFERENCE IN METAL UPTAKE FROM SLUDGE-AMENDED SOIL

Cultivars (also referred to as varieties, genotypes, selections or strains) within a crop species vary significantly in

Table 29. Relative uptake of trace elements to tissues of forage crops
(dry weight basis) (Carlton-Smith and Davis, 1983).

Crop	Cultivar	% of crop with highest concentration				
		Cd	Cu	Mo	Ni	Zn
Agrostis tenuis	'Goginan'	19	49	38	73	36
Agrostis tenuis	'Parys'	12	55	50	85	44
Dactylis glomerata	'S26'	39	100	—	57	17
Dactylis glomerata	'S37'	15	57	—	33	10
Dactylis glomerata	'S143'	12	47	—	30	11
Festuca arundinacea	'S170'	52	76	23	63	32
Festuca pratensis	'S215'	37	78	—	91	32
Festuca rubra	'Merlin'	10	51	—	34	19
Festuca rubra	'S59'	29	51	—	47	22
Lolium multiflorum	'Aubade'	15	68	29	44	24
Lolium multiflorum	'RVP'	17	72	28	55	25
Lolium multiflorum	'S22'	15	73	29	46	26
Lolium multiflorum	'Sabalan'	15	67	25	59	21
Lolium multiflorum	'Sabrina'	11	89	29	60	26
Lolium perenne	'Cropper'	16	86	37	80	31
Lolium perenne	'Melle'	15	86	27	92	27
Lolium perenne	'S23'	14	86	33	68	30
Lolium perenne	'S24'	14	90	29	100	30
Lolium perenne	'S321'	19	90	29	80	30
Lolium perenne	'Talbot'	14	89	35	82	32
Phleum pratense	'S48'	31	96	25	73	30
Phleum pratense	'S51'	31	95	24	73	28
Phleum pratense	'S352'	32	88	21	80	33
Avena sativa	'Trafalgar'	9	29	16	42	8
Hordeum sativa	'Julia'	9	48	16	16	15
Triticum aestivum	'Sappo'	7	38	16	17	10
Zea mays	'Caldera'	13	30	11	16	10
Zea mays	'Maris Carmine'	32	31	12	16	14
Medicago sativa	'Europe'	14	35	34	31	11
Trifolium pratense	'Hungarapoly'	7	53	62	44	14
Trifolium pratense	'S123'	7	57	62	48	16
Trifolium repens	'Kent Wild White'	8	43	100	34	18
Trifolium repens	'S100'	5	51	79	36	20
Trifolium repens	'S184'	5	38	71	40	21
Brassica oleracea	'Maris Kestrel'	16	21	41	30	13
Brassica rapa	'The Bruce'	50	35	38	33	19
Beta vulgaris	'Sharpes Monobeet'	100	66	19	57	83
Beta vulgaris	'Yellow Globe'	95	81	26	57	100
Highest concentration among crops (mg/kg dry weight)		1.41	15	14	1.4	417

Concentration in plant shoots normalized across control and 2 sludge
treatments. For the crop with the maximum normalized concentration,
(100) = the listed mg element/kg dry matter.

uptake of sludge-applied trace elements. Cultivar variation in
Cd and other element uptake was evaluated because in agronomic
management unrelated to sludge use (Foy et al., 1978) this source
of variation had been found to be important in correction of
trace element deficiencies (Fe, Cu, Mn, Zn) and in tolerance of
plants to trace element toxicity (Al, Mn, Zn). The expected
benefits from cultivar difference in tolerance or uptake of
sludge-borne trace elements include: (1) the ability to select
relatively metal-tolerant or metal-excluder (non-accumulator)
cultivars for use in management of designed sludge farms; (2)
determination of whether cultivar differences are great enough to
require adjustment of dietary element risk assessments; and (3)
identification of cultivars which could be used to reduce
background levels of element ingestion from the general food
supply.

At this time, a few crops have been studied under the conditions which generate results considered reliable for evaluation
of the long-term effects of sludge-borne trace elements. Cadmium
uptake by cultivars of corn, soybean, and lettuce have been

Table 30. Relative Cd concentration in crops grown on naturally Cd
rich Salinas Valley soils (Burau, 1980).*

Crop	$\frac{mg\ Cd/kg\ wet\ weight}{mg\ Cd/kg\ dry\ soil}$	% dry weight†	$\frac{mg\ Cd/kg\ dry\ weight}{mg\ Cd/kg\ soil}$	Relative uptake
Spinach	0.70	9.3	7.5	280
Endive	0.24	6.9	3.5	130
Lettuce, leaf	0.16	6.0	2.7	100
Lettuce, romaine	0.16	6.0	2.7	100
Lettuce, head	0.07	4.5	1.6	59
Chili pepper	0.10	8.0	1.2	44
Carrots	0.13	11.8	1.1	41
Artichokes	0.14	13.5	1.0	37
Potatoes	0.09	20.2	0.45	17
Garlic	0.17	38.7	0.44	16
Sweet corn	0.11	27.3	0.40	15
Cucumber	0.02	4.9	0.41	15
Squash, zucchini	0.02	5.4	0.37	14
Red beets	0.04	12.7	0.31	11
Onions	0.03	10.9	0.28	10
Cauliflower	0.02	9.0	0.22	8
Parsley	0.03	14.9	0.20	7
Tomatoes	0.01	6.5	0.15	6
Broccoli	0.01	10.9	0.092	3
Beans, white	0.08	89.1	0.090	3

*Soils were between pH 6–8 and contained 1–10 mg Cd/kg, and about 1.5%
organic matter.

†From Adams (1975).

studied in appreciable detail. Carrot, wheat, and some forage
species have been studied, but to a lesser extent. Other studies
are needed, particularly for crops which strongly absorb or
exclude particular trace elements. As a general rule, cultivars
have been found to vary by at least 2- to 5-fold from lowest to
highest uptake response. However, a 30-fold variation was found
in corn inbreds.

An extensive characterization of relative corn cultivar
variation in uptake of a sludge-applied trace elements was
reported by Hinesly et al. (1978, 1982a). Uptake of Cd and Zn
by 20 corn inbreds grown on long-term sludge-amended soils in the
field were reported. These soils provided substantially
increased plant-available Cd and Zn. The relative Cd and Zn concentration in the cultivars were recalculated as the geometric
mean for three sludge rates. Leaf Cd ranged from 0.88 mg Cd/kg
dry weight in inbred R805 to 30.3 mg Cd/kg dry weight in inbred
B37. Grain Cd ranged from 0.05 mg Cd/kg dry weight in inbred H96
to 1.81 in inbred B37. The grain and leaf Cd concentrations were
highly correlated as were the ranks among inbreds of grain Cd
concentration and leaf Cd concentration. However, the grain Cd
to leaf Cd concentration ratio of an inbred varied from 0.018 to
0.104. This wide range in grain Cd to leaf Cd ratio indicates
that one should not base a breeding program to lower grain Cd
concentrations only on measuring seedling leaf Cd concentrations,
nor should one use grain Cd results alone to select for Cd-
excluder silage corn cultivars. On these same field plots, corn
leaf Zn ranged from 44.2 to 152 mg/kg dry weight and grain Zn
varied from 31.5 to 58.4 mg/kg. Although ranks and concentration

of both Cd and Zn in corn inbreds were each well correlated over
sludge rates, concentrations of Cd and Zn were not significantly
correlated, nor were cultivar ranks for Cd and Zn concentrations
significantly correlated. Thus, mechanisms controlling uptake or
translocation to corn grain may vary for Zn and Cd.

Hinesly et al. (1978) then used the results from study of
corn inbreds to make selected corn single-cross hybrids, and
evaluated Cd and Zn uptake to leaves and grain, Cd and Zn distribution in milling fractions (Hinesly et al., 1979), and Cd
bioavailability to laying hens (Hinesly et al., 1985). Single-
cross corn hybrids were prepared which accumulated low or high
levels of Cd into grain. Significant reduction in food-chain
transfer of Cd is possible by selecting for Cd-excluder corn
hybrids.

Bache et al. (1981) compared the accumulation of Cd and
other elements by 8 corn hybrids adapted to upper New York State.
In a greenhouse study with freshly applied air-dried sludge, a
6-fold range in grain Cd and 2-fold range in stover Cd were found
among the hybrids tested. The possible range of response is
probably fairly widely expressed in the wide range of corn
hybrids of commercial importance in the U.S., but this range is
not nearly as wide as seen for corn inbreds.

The Cd uptake to shoots of seedling plants of soybean cultivars grown on sludge-amended soil showed about a 4-fold range for
10 cultivars (Boggess et al., 1978). The investigators also
reported Cd uptake from soils amended with Cd salts; although
relative Cd concentrations among cultivars were similar, they
were not related closely enough to rely on salt-Cd results to
predict cultivar uptake on sludge-amended soils. Hinesly (1986,
Personal communication) found a smaller range in concentration of
Cd in grain of soybean cultivars grown to maturity in the field,
and the ranks of grain Cd concentration among 3 cultivars was
similar to the findings of Boggess et al. (1978) for seedling
soybean shoots.

Following a study with 3 lettuce cultivars (Harrison, 1986a),
Yuran and Harrison (1986) studied relative Cd uptake by 60 lettuce genotypes. They grew the 60 genotypes in 1 soil type, with
1 rate (90 mt/ha/yr) of 2 sludges (12 and 57 mg Cd/kg dry
sludge), for 2 years. Soil Cd in year 2 was 0.09, 1.1, and 9.0
kg Cd/ha for the control, lower Cd sludge, and higher Cd sludge,
respectively. Soil pH was adjusted to 5.7 by S addition. Some
genotypes varied in response between years, while others did not
(genetic studies are being conducted). In 1984, the geometric
mean response for all genotypes on the untreated soil was 1.7 mg
Cd/kg dry weight (range 1.2-2.5), 2.6 mg Cd/kg for the lower Cd
sludge (range 1.4-5.5), and 5.6 mg Cd/kg for the higher Cd sludge
(range 3.7-11.9). Thus, the field-determined variation in lettuce foliar Cd concentration was only about 3-fold.

Other less intensive studies with sludge-amended soils confirm this relatively narrow range (Giordano et al., 1979; Davis
and Carlton-Smith, 1980 [Table 27]; Feder et al., 1980 [see CAST,
1980, Table 8] for leafy type lettuces. Head lettuce had
approximately half the Cd concentration found in leafy lettuces
grown in Salinas Valley (Burau, 1980) (Table 30). Chaney and

Munns (1980, unpublished) tested the effect of sludge source and soil pH on Cd uptake by 6 lettuce cultivars. They also found cultivar differences in Cd uptake were small. Relative cultivar response was similar on lower and higher pH soils, but the range was much narrower on limed soils. Davies and Lewis (1985) and Crews and Davies (1985) compared trace element concentration in edible tissues of lettuce cultivars grown on metal rich soils contaminated with mine wastes in Great Britain. They found that relative cultivar response was similar on different soils, and the range of lettuce Cd concentration was about 2- to 3-fold on the several soils.

The Cd concentration in potato tubers did not vary significantly among 6 cultivars grown on a metal-rich soil at a long-term sludge utilization farm (Harris et al., 1981). The soil was pH 6.6 and contained 19.6 mg Cd/kg, while the mean Cd level for the washed unpeeled potato tubers was 0.28 mg Cd/kg dry weight [slightly greater than background Cd level in U.S. potatoes, 0.165 mg/kg dry weight (Wolnik et al., 1983b)].

Meyer et al. (1982) found substantial differences among wheat types grown on U.S. soils containing background Cd levels. Durum type cultivars contained 0.140 mg Cd/kg dry grain, while soft red spring, soft red winter, hard red spring, and white wheat cultivars contained only 0.044 mg Cd/kg. Grain Cd was not significantly correlated with soil total Cd across all wheat cultivars, but was correlated if wheat types grown on similar soils were examined. Additional information has been provided by Hinesly (1986, Personal communication) on grain Cd in different cultivars of wheat grown on sludge-amended soil. 'Beau' grain contained 3.4 mg Cd/kg, while 'Argee' contained only 2.4 mg Cd/kg (strongly acidic soil, pH 5.5; 0.1 M HCl-extractable soil Cd about 33 mg/kg).

In a cooperative field trial on sludge metal availability, 4 barley cultivars from different regions of the U.S. were compared in a greenhouse experiment (Chang et al., 1982); no significant differences were found among the cultivars in Cd or Zn uptake to leaves or grain from plants grown on sludge-treated soil.

The effect of sewage sludge and carrot genotype on Cd accumulation in edible carrot roots was reported by Harrison (1986b). Two sludges were applied to 1 soil in 3 bed configurations. The mean Cd level was 0.38 mg Cd/kg for the control carrots, 0.50 mg Cd/kg for carrots grown on the lower Cd level sludge, and 0.77 mg Cd/kg for the higher Cd sludge. Genotypes differed less than 2-fold in Cd accumulation. Differences among hybrid selections were significant for Cd, Zn, and other elements, although not all elements were increased due to sludge application.

ESTIMATING MAXIMUM ALLOWABLE SOIL Cd LOADING
BASED ON PREDICTED INCREASE IN DIETARY Cd

Several methods have been used in different nations and at different times to estimate the maximum cumulative Cd application which protects the health of individuals (Dean and Suess, 1985). This is a very complex issue, as has been noted by Ryan et al. (1982) and Logan and Chaney (1983). The analysis given in Ryan et al. (1982) was considered when the US EPA proposed the existing regulations on land application of sludge in 1979 (Environmental Protection Agency, 1979a). A background document (Environmental Protection Agency, 1979b) reporting the scientific basis for the regulations was released at the time the interim final regulations were published.

Based on FDA dietary Cd intake estimates (36 µg Cd/day) and WHO/FAO recommendations for maximum tolerable weekly Cd intake (52-71 µg Cd/day), EPA (1979a) concluded that sludge could safely add no more than 30 µg Cd/day to an individual's diet. The high-risk or high-exposure individual was to be protected by the regulation: "That high-risk situation is one where an individual receives 50% of his vegetable diet from sludge-amended soils for a period of 40 to 50 years." The U.S. EPA recognized the strong effect of soil pH on Cd uptake by crops. For soils with low background pH, it was considered likely that soil pH would fall (from the pH 6.5 required during the permitted period of sludge application) to background soil pH. Thus, data from crops grown on acidic sludge-amended soils were used to estimate the relative Cd uptake by different food groups. The background document cites work by Dowdy and Larsen (1975), Giordano and Mays (1977), Chang et al. (1978), and a pot study by Furr et al. (1976b). EPA calculated the increase above control, relative to that for lettuce, for each sludge application rate. The relative increases were averaged across rates; radish and carrot were averaged to obtain "root vegetables", and pea fruits and pea pods were averaged to obtain "legume vegetables." Lettuce represented "leafy vegetables" and tomato represented "garden fruits."

Table 31 summarizes the presumed relative increased Cd uptake by crops in the relevant FDA food classes, and daily food intakes for the teenage male diet model used by EPA in 1979 (Environmental Protection Agency, 1979b). If one multiplies food intakes (g dry/day, column B) times relative increased Cd uptake (column D), one obtains relative increased daily Cd intakes (column E). Thus, if lettuce is increased by 1 mg Cd/kg dry weight, garden foods are increased by 7.90 µg Cd/day for 100%, or 3.95 µg/day for 50% of garden foods grown in acidic sludge-amended garden for 40-50 years. EPA judged that strongly acidic soils (pH 5.4 to 6.2) with 5 kg sludge-applied Cd/ha would not cause greater than 30 µg Cd increase/day, although very acid soils (4.9) caused larger increases.

Several groups evaluated the 1979 regulations, and other research provided new information on and a better understanding of Cd transfer and food consumption. By 1980, when U.S. EPA's Office of Solid Waste was preparing regulations under The Clean Water Act Section 405d, it was clear that average adult dietary intake data rather than teen-aged male dietary intakes should be used. Pennington (1983) provided an early draft of her results, and these were summarized into food groups by Flynn at EPA (1986,

Table 31. Cadmium exposure model from the 1979 Environmental Protection Agency sludge application regulation and background document (EPA, 1979a, 1979b), and the 1981 draft background document. Table shows intakes of FDA food classes by the hypothetical teenaged male diet model (1979) or average adult diet model (1981), and relative Cd uptake by food groups (EPA, 1979b).

Food Group	A	B	C % Dry Weight	D Relative Increased Cd Uptake	E = (B x D) Relative Daily Cd
	Food Intake				
	g wet/day	g dry/day			µg Cd/day
1979 Diet Model					
Leafy vegetables	55	4.95	9	1.00	4.95
Potatoes	183	43.9	24	0.02	0.88
Root vegetables	33	2.64	8	0.23	0.61
Legume vegetables	69	13.1	19	0.04	0.52
Garden fruits	69	5.52	8	0.17	0.94
					7.90

If 100% of garden foods diet were grown on acidic sludged land, diet would be increased 7.90 µg Cd/day when lettuce increased by 1 µg Cd/g dry weight. If 50% of garden vegetables, diet increases by 3.95 µg/day when lettuce increases 1 µg Cd/g dry weight.

1981 Diet Model					
Leafy vegetables	26	2.34	9	1.00	2.34
Potatoes	64	15.36	24	0.02	0.307
Root vegetables	13	1.04	8	0.23	0.239
Legume vegetables	38	7.22	19	0.04	0.289
Garden fruits	60	3.60	6	0.17	0.612
					3.79

If 100% of garden foods diet were grown on acidic sludged land, diet would be increased 3.79 µg Cd/day when leafy vegetables increased by 1 µg Cd/g dry weight. If 50%, diet increases by 1.90 µg/day; and if 33%, 1.26 µg Cd/day.

Personal communication). Leafy vegetables included lettuce, spinach, collards, cabbage, coleslaw, and sauerkraut. Potatoes included french fries, mashed, baked, boiled, scalloped, and sweet potatoes, and potato chips. Root vegetables included carrots, onions, beets, radishes, onion rings, mushrooms, and mixed vegetable. Legume vegetables included pinto, lima, navy, green (snap), and red beans, pork and beans, cowpeas, peas, peanuts, and peanut butter. Garden fruits included cucumber, pickles, tomatoes, tomato sauce and juice, catsup, cream tomato soup, squash, and vegetable soups; broccoli, celery, asparagus, and cauliflower were included here by Flynn because they have Cd response more similar to garden fruits than leafy vegetables, which they had had been classified in the 1979 teen-age male diet model.

The 36 µg Cd/day average intake (from FDA) was subtracted from the 71 µg/day WHO/FAO limit yielding 35 µg Cd/day allowed increase. Using the average adult rather than the teenage male dietary intakes reduced the predicted increase from 7.90 to 3.79 µg Cd/day (for 100% of garden vegetable foods grown on acidic sludge amended soil) when lettuce is increased by 1 µg Cd/day dry weight (Table 31). If one divides 35 µg Cd/day by 3.79 µg increased dietary Cd (per 1 µg Cd increase/g dry lettuce), one finds leafy vegetables could safely reach 9.23 µg Cd increase/g for 100% of diet; or 18.5 µg Cd increase/g lettuce for 50% of diet; or 27.7 µg Cd increase/g for 33% garden foods diet grown on acidic sludge-amended soils.

EPA then attempted to connect these leafy vegetable Cd increases to cumulative soil Cd loadings. Results for cumulative soil loading versus leafy vegetables were separated into acidic (pH 5.4-6.2) [Y(lettuce Cd) = 0.48X(kg Cd/ha) + 5.6; R^2 = 0.11] and very acidic (pH < 5.3) [Y(lettuce Cd) = 8.1 X(kg Cd/ha) − 1.1; R^2 = 0.50]; each had low R^2. They then tested the annual soil Cd loading results for acidic soils and got a better correlation [Y(lettuce(Cd)) = 1.24X(kg Cd/ha) + 0.12; R^2 = 0.84]. Thus, using this equation, and 33% of garden foods, one calculates that soil could contain 22.2 kg added Cd/ha, or about 10 mg/kg.

The effect of soil pH on relative increased Cd uptake by garden crops is demonstrated by the data in Table 32. Strongly

Table 32.   Effect of soil pH on relative increase above control of Cd in edible crop tissues (Chaney, 1985. Personal communication).

Crop	Increased Crop Cd		Relative Increase	
	Acidic	Limed	Acidic	Limed
	―mg Cd/kg dry―			
Lettuce	29.3	5.73	1.00	1.00
Carrots	2.15	1.48	0.073	0.26
Potatoes	1.17	1.02	0.040	0.18
Peanuts	0.54	0.41	0.018	0.072

Sludge containing 210 mg Cd/kg applied at 50 and 100 mt/ha in summer, 1978. Carrot and lettuce results from 1979; potato and peanut results from 1980.

acidic soil pH causes much greater Cd uptake than near neutral soil pH, especially for the Cd-accumulating leafy vegetables. The relative increased Cd uptake is greater for carrot, potato, and peanut at the higher soil pH. Because high crop uptake at acidic soil pH is required for appreciable risk from nearly all sludges, relative increased Cd uptake for acidic soils should be used for risk assessment. However, this source of variation should be considered in evaluating different sources of data on relative Cd uptake by crops.

Relative increased Cd uptake by food groups from studies reported above are summarized in Table 33. Results varied among studies due to soil pH and perhaps due to sludge properties and crop cultivar. The results from Burau (1980) were not "increased Cd uptake", but the slope from actual Cd uptakes; further, high pH soils were sampled in his study. Potato, legume vegetables, and garden fruits have low relative increased Cd uptakes in all studies (ignoring potato in the Burau data set). The freshly

Table 33. Comparison of relative increased Cd uptake by food groups based
on different data sources summarized above.

Food Group	Reference[†]					
	EPA 1979	Davis (28)	Dowdy Larson 1975	Giordano et al. 1979	Burau (30)	Chaney Acidic (32)
Leafy Vegetables (lettuce)	1.00	1.00	1.00	1.00	1.00	1.00
Potato	0.02	0.020	0.052	0.00	0.17	0.040
Root Vegetables	0.23	0.07	0.36	0.37	0.21	0.073
Legume Vegetables	0.04	0.01	0.022	0.017	0.03	0.014
Garden Fruits	0.17	0.020	0.15	0.18	0.12	—

[†]Numbers in parentheses indicate table in text where data are contained.

added sludge in the Dowdy and Larson (1975) study may have
altered relative crop uptake of Cd. More field data from
strongly acidic long-term sludge-amended soils are needed,
especially for responsive conditions, and enough examples of the
food groups to provide useful data. Potato, high and low-Cd
accumulative leafy and root vegetables and garden fruits, and
green beans or peas should be grown. Some questions remain about
use of high Cd sludges to provide responsive conditions which
allow statistically significant differences in Cd uptake among
crops. Although the relative increased Cd uptake may appear much
higher for low Cd concentration sludges, these data are not relevant to soil Cd conditions in which increased crop Cd might be
important to health. The precision of this approach is not as
good as desired, but the estimation of Cd transfers to the food-
chain is the best available with present data. The true worst-
case is more precisely evaluated by the proposed method than is
the low sludge Cd case which provides a high margin of safety.
Until the method for assessing Cd transfer to the food-chain is
refined, allowing somewhat lower sludge Cd concentrations than
the calculated maximum amount would provide an increased safety
factor.

The next step is to combine the best data on food intakes and
on relative increased Cd uptake. Upon examination of the food
intake data, several differences were found between the final
published data of Pennington (1983) and the adult food intakes
prepared by Flynn. Table 34 shows the average adult intake of
foods (averaged over male and female, age groups 14-16 years,
25-30 years, and 60-65 years), based on the published data of
Pennington (1983). Dry matter percentage for each food were
obtained from Adams (1975). Further, because food groups contained crops with large differences in relative Cd uptake, leafy
and root vegetables and garden fruits were separated into higher
and lower Cd accumulating groups. Sweetcorn, strawberries, and
melons were added to garden fruits (not included by Flynn).
Lastly, 4 vegetables (broccoli, cauliflower, asparagus, and
celery) were moved back to the low Cd-accumulating leafy vegetable group. These results are summarized in the revised food

Table 34.   Average adult daily intakes of foods aggregated into food groups
on wet weight and dry weight basis. Average wet g/day food intakes obtained from Pennington (1983); six male and female
diets, for ages 14-65, were averaged. Converted to dry weight
using data from Adams (1975).

Pennington Data		Adams Data	Adult Food Intake	
Food	g wet/d	% dry wt.	g dry/day	% dry
Leafy Vegetables – High Cd Uptake:				
Lettuce	19.231	4.5	0.865	
Spinach	0.816	8.6	0.070	
Spinach	2.329	8.0	0.186	
	22.376		1.121	5.0
Leafy Vegetables – Low Cd Uptake:				
Collards	1.715	10.4	0.178	
Cabbage	2.849	6.1	0.174	
Coleslaw	2.530	19.4	0.491	
Sauerkraut	0.939	7.2	0.068	
Broccoli	2.403	8.7	0.209	
Celery	0.922	5.9	0.054	
Asparagus	0.836	6.4	0.054	
Cauliflower	0.772	7.2	0.056	
	12.966		1.284	9.9
Potatoes:				
French fries	20.026	55.3	11.074	
Mashed	16.232	20.7	3.360	
Boiled	12.202	20.2	2.465	
Baked	6.859	24.9	1.708	
Chips	2.963	98.2	2.901	
Scalloped	5.941	28.9	1.717	
Sweet	1.541	36.3	0.559	
Sweet	0.674	40.0	0.270	
	66.438		24,054	36.2
Root Vegetables – High Cd Uptake:				
Carrots	3.401	17.8	0.605	17.8
Root Vegetables – Low Cd Uptake:				
Onions	2.473	10.9	0.270	
Mixed veg.	5.154	17.4	0.897	
Mushroom	0.787	9.6	0.076	
Redbeets	1.069	10.7	0.114	
Radish	0.402	5.5	0.022	
Onion rings	0.710	8.2	0.058	
	10.595		1.437	13.6
Legume Vegetables:				
Snap beans	2.924	7.6	0.222	
Snap beans	6.172	8.1	0.500	
Cowpeas	2.402	28.2	0.677	
Lima beans	0.870	35.9	0.313	
Lima beans	1.301	26.5	0.345	
Navy beans	1.567	31.0	0.486	
Red beans	1.748	31.0	0.542	
Peas	5.568	23.0	1.281	
Peas	1.533	17.9	0.274	
Peanut butter	2.269	98.3	2.230	
Peanuts	0.896	98.2	0.880	
Pinto beans	7.589	31.0	2.353	
Pork & beans	7.552	33.6	2.537	
	42.391		12.640	29.8
Garden Fruits – High Cd Uptake:				
Tomatoes	16.186	6.5	1.053	
Tomato juice	3.658	6.4	0.234	
Tomato sauce	2.108	13.0	0.274	
Tomatoes	0.885	6.3	0.056	
Tomato soup	9.464	9.5	0.899	
Catsup	2.396	31.2	0.748	
Green pepper	0.830	6.6	0.055	
	35,437		3.319	9.4
Garden Fruits – Low Cd Uptake:				
Cucumber	2.735	4.3	0.118	
Pickles	1.324	6.7	0.089	
Summer squash	2.190	4.5	0.099	
Winter squash	1.207	11.2	0.135	
Watermelon	4.748	7.4	0.351	
Canteloupe	4.416	8.8	0.389	
Strawberries	2.508	10.1	0.253	
Sweet corn	6.574	25.9	1.703	
Sweet corn	3.816	25.1	0.958	
Sweet corn	2.431	23.7	0.576	
	31.949		4.671	14.6

groups in Table 35. Thus, the revised sum of increased Cd intakes, due to consumption of 100% of garden foods (grown on acidic
sludge-amended soil) for the average adult is 2.20 µg Cd/day when
lettuce is increased by 1 mg Cd/kg dry weight. If one assumes
that only 50% of one's garden foods are grown on acidic sludge
amended soils, dietary Cd increases only 1.10 µg/day when lettuce
Cd is increased by 1 mg/kg dry weight.

Table 35. Food group aggregation of food intake results from Pennington
(1983). Data for six age-x-sex groups (ages 14-65) were
averaged; wet weight conversion to dry weight conducted on
individual food basis using data from Adams (1975). Foods
from the Pennington lists were the same as listed by Flynn
except stalk vegetables were moved to leafy vegetables -
low category, and sweet corn and melons were added to garden
fruits - low.

Food Group[†]	Food Intakes wet g/d	dry g/d	% Dry Weight	Relative Increased Cd Uptake	Relative Daily Cd Intake
Leafy Vegetables-High	22.376	1.121	5.0	1.00	1.121
Leafy Vegetables-Low	12.966	1.284	9.9	0.13	0.167
Potatoes	66.438	24.054	36.2	0.020	0.481
Root Vegetables-High	3.401	0.605	17.8	0.20	0.121
Root Vegetables-Low	10.595	1.437	13.6	0.052	0.075
Legume Vegetables	42.391	12.640	29.8	0.010	0.126
Garden Fruits-High	35.537	3.319	9.3	0.020	0.066
Garden Fruits-Low	31.949	4.671	14.6	0.010	0.047
					2.20

[†]Leafy Vegetables - High includes lettuce and spinach.
Leafy Vegetables - Low includes cabbage, kale, broccoli, etc.
Root Vegetables - High includes carrots.
Root Vegetables - Low includes radish, turnip, beet, onion, and leek.
Garden Fruits - High includes tomato products and pepper.
Garden Fruits - Low includes cucurbits, sweet corn, and strawberries.

Table 36 shows a comparison of the food intake, relative Cd
uptake, and relative Cd intakes for the three estimates discussed
above. The food intakes used in the model have changed substantially due to change in age group and diet survey relied upon.
The greatest declines were in amount of potato and leafy vegetables ingested; these changed due to change in dietary patterns
during 20 years, and use of actual survey results rather than the
teen-aged male diet with other foods added to balance the diet.
Separating food groups into high and low Cd accumulating sub-
groups further lowered the estimated increase in dietary Cd, but
using the final Pennington data set with correct dry weight per-
centages increased the food intake estimates compared to those of
Flynn. The final workgroup estimate is less than 1/3 as high as
the 1979 estimate, indicating that the worst-case acidic garden
food-chain model has less than 1/3 the risk estimated in the 1979
regulation development process.
    To estimate the worst-case Cd transfer to garden crops for a
particular sludge and soil, one multiplies 2.20 µg Cd/day (for
100% of garden foods) times the increased Cd in lettuce above the
background when the plateau is reached at increasing sludge rate,
and at strongly acidic soil pH. The data of Chaney (Figure 2)

Table 36. Comparison of food intakes, relative increased Cd uptake, and estimated increased dietary Cd in the EPA (1979b), Flynn (1981), and present document.

Food Group	EPA, 1979	Flynn, 1981	Present
		g dry weight/day	
Food intakes:			
Leafy vegetables	4.95	2.34	2.405
Potato	43.9	15.36	24.063
Root vegetables	2.64	1.04	2.042
Legume Vegetables	13.1	7.22	12.640
Garden fruits	5.52	3.69	7.990
Relative food group increased Cd uptake (lettuce = 1.00):			
Leafy vegetables	1.00	1.00	0.536
Potato	0.02	0.02	0.02
Root vegetables	0.23	0.23	0.096
Legume Vegetables	0.04	0.04	0.010
Garden fruits	0.17	0.17	0.014
Estimated relative food group increased Cd intake:			
Leafy vegetables	4.95	2.34	1.288
Potato	0.88	0.307	0.481
Root vegetables	0.61	0.239	0.196
Legume Vegetables	0.52	0.289	0.126
Garden fruits	0.94	0.612	0.113
Increased dietary Cd (μg/day) if lettuce increased 1 mg Cd/kg dry:			
	7.90	3.79	2.20

Present estimate differs from previous estimates by separating leafy, root, and legume vegetable and garden fruits into high and low Cd-accumulating crop sub-groups. The overall food group relative increased dietary Cd intake was calculated using the increased food and Cd intake from high and low sub-groups (from Table 35).

were subjected to non-linear regression analysis to estimate the plateau in lettuce Cd for the acidic heat-treated sludge. The estimated increase above control was 0.93 mg Cd/kg dry lettuce (upper 95% confidence interval was 1.21 mg Cd/kg based on non-linear regression using 3 replications for 7 crop years) in strongly acidic soil (the worst-case condition), and 0.29 mg/kg (upper confidence interval was 0.40) in limed soil when low Cd, Fe-rich sludge was applied. These convert to about 1.02 (upper 95% confidence interval = 1.33) and 0.32 (upper 95% C.I. = 0.44) μg Cd/day increased Cd intake (assuming 50% of dietary garden vegetables are grown on sludge-amended soils).

TRANSFER OF SLUDGE-APPLIED TRACE ELEMENTS TO ANIMALS BY DIRECT INGESTION OF SLUDGE OR SLUDGE-AMENDED SOIL

Sludge is often utilized on pastures because the land is accessible intermittently during the growing season. Spray-applied sludge can adhere to forage crops and allow direct ingestion of the sludge particulates. Alternatively, surface-applied sludge materials become part of the organic-rich "thatch layer" on the soil surface. Grazing livestock inadvertently ingest surface soil. These processes can allow much more direct exposure to organic and inorganic constituents in sludge than can occur via plant uptake.

Several studies have evaluated crop and sludge factors in adherence of spray-applied sludge to forage crops. Increased sludge application rate (within practical rates) increased sludge on the forage (Chaney and Lloyd, 1979). If sludge dried on the crop, it remained largely there regardless of rainfall. Jones et al. (1979) found that sludge could be washed off immediately after application, but not after it dried. They further showed that sludges with higher solids content caused higher increased sludge adherence, perhaps because of viscosity.

Chaney and Lloyd (1986, Personal communication) completed 2 further studies on sludge adherence to evaluate crop species and sludge properties effects. In the first study, they compared 2 sludges on 2 crops (Table 37). Because the sludge applied by

Table 37.   Effect of sludge source, and time after sludge application on sludge adherence to tall fescue and orchardgrass (Chaney and Lloyd, 1986, Personal communication).

Sludge	% Solids	Crop	\multicolumn{4}{c}{Days After Sludge Application}			
			0	7	14	20
			\multicolumn{4}{c}{-------% sludge in/on forages-------}			
City 1	2.0	Tall Fescue	3.0	3.1	1.5	0.82
		Orchardgrass	7.4	4.4	2.8	0.90
City 23	7.6	Tall Fescue	10.2	6.2	3.1	2.7
		Orchardgrass	11.9	6.9	2.1	1.0

Forages were not clipped; sludge applied at 94 m³/ha using watering cans. Sludge content calculated based on increased levels of 6 elements above levels present in unsprayed control forage samples.

Anaerobically digested sludge from City 1 contained (in mg/kg dry solids): Zn, 3030; Cd, 54.9; Pb, 495; Cu, 665; Ni, 68; Fe, 11,000.

Anaerobically digested sludge from City 23 contained (in mg/kg dry solids): Zn, 750; Cd, 7.2; Pb, 170; Cu, 195; Ni, 27; Fe, 123,000.

Chaney and Lloyd (1979) was very high in Fe, they used a second sludge with more normal Fe level. The 2 sludges differed strongly in solids content, and much more sludge adhered at 7.6% solids than at 2.0% solids. This study was done during the active growing season, and growth rapidly diluted the sludge content of forage. Although orchardgrass had slightly higher sludge adherence initially, either more rapid growth diluted adhering sludge on orchardgrass than on the tall fescue, or sludges did not adhere as long on orchardgrass.

In the second study, 1 sludge was applied at 2 rates to unclipped or clipped stands of 5 forage crop species, and forage samples were harvested at 6 times until after the normal harvest age of the clipped areas (Table 38). Statistical analysis showed that the harvest date was the most significant treatment variable, although the highest significant interaction term was clipping-x-species-x-harvest date. Alfalfa had the highest sludge content initially, but the lowest by the final harvest. Bluegrass had the second highest content initially, and retained the highest content during growth. Alfalfa's growth pattern is different from that of the grasses; the new growth of cut alfalfa comes from the highest axillary nodes, and all new growth emerges well above the thatch layer and contains no sludge. Treated parts of the grasses rise as new growth occurs at their base;

Table 38.   Effect of forage crop species, clipping crop before sludge application, and time after application on adherence of spray-applied fluid sludges to five forage crop species (Chaney and Lloyd, 1986, Personal communication).

Crop	Harvest Date	Unclipped	Clipped
		——% sludge in/on forage——	
Tall Fescue	0	3.69 d-i	4.44 cd
'Kentucky 31'	7	3.02 f-l	3.77 c-h
	14	2.79 h-n	2.61 k-o
	28	1.89 m-q	1.19 q-v
	43	2.85 h-m	0.53 u-x
	70	1.01 q-w	0.30 vwx
Orchard grass	0	2.68 i-o	4.53 cd
'Potomac'	7	2.27 l-p	3.67 d-i
	14	2.34 k-p	1.67 o-s
	28	1.61 p-t	0.60 t-x
	43	1.32 p-v	0.40 vwx
	70	1.11 q-x	0.13 wx
Kentucky bluegrass	0	6.36 b	5.68 b
'Merion'	7	4.18 cde	3.99 c-f
	14	4.73 c	3.33 e-k
	28	4.37 cd	2.97 g-l
	43	2.64 j-o	1.32 p-v
	70	1.81 n-q	0.52 u-x
Smooth bromegrass	0	3.63 d-j	4.26 cde
'Saratoga'	7	3.29 e-l	4.31 cde
	14	2.75 h-n	3.97 c-g
	28	1.45 p-u	1.70 o-r
	43	1.56 p-t	0.62 t-x
	70	0.77 r-x	0.45 u-x
Alfalfa	0	8.48 a	5.98 b
'Saranac'	7	6.17 b	4.30 cde
	14	3.67 d-i	1.82 n-q
	28	1.21 q-v	0.51 u-x
	43	0.66 s-x	0.13 wx
	70	0.09 wx	-0.05 x

Forages were established in spring, 1976 on methyl-bromide treated field plots. After establishment, sludge was applied on May 11, 1977, at 51 and 103 m³/ha, to unclipped and clipped (to 10-15 cm as recommended for species, with clippings removed) crops in three replications. Forage was harvested to 5 cm after 0-70 days growth. Normal harvest of the clipped forage would have occurred about day 43.

Sludge was 1.4% solids and contained (in mg/kg dry solids): Zn, 1140; Cu, 432; Pb, 394; and Fe, 36,000. Sludge content was estimated by increased levels of Zn, Cu, Pb, and Fe in sprayed forage. The species—clipping—harvest date was the highest significant interaction in ANOVA. Sludge content results followed by the same letter were not significantly different (at P < 0.05) according to the Duncan Multiple Range Test.

new leaves emerge at the base and grow through the thatch layer. Thus, alfalfa and other dicotyledonous forage species are unlikely to allow much sludge transfer to the food chain under good management practices. Bluegrass forms very tight bunches, and sludge particles are trapped within the harvestable portion of the crop. These studies, taken together, indicate that present advice of "clip before spray application, avoid high solids content sludges, and wait for normal regrowth of the crop before harvest or grazing" continues to reflect research findings. Bluegrass is a species particularly inappropriate for spray

application of sewage sludge. Application rate, although statistically significant, was the least important factor studied.
   Although forages can reach 15-30% sludge (dry matter basis)
immediately after application, Environmental Protection Agency
regulations (EPA, 1979a) now require a 30-day waiting period
before grazing; users are advised to apply sludge to well grazed
or clipped forages. This requirement reduces initial adherence
and growth rapidly dilutes the adhering sludge. Decker et al.
(1980) and Bertrand et al. (1981) found sludge comprised only
2-3% of the dry diet (based on forage and feces analyses) (Table
39) in practical grazing management. Injection of sludge in the

Table 39. Adherence of spray-applied liquid sewage sludge to tall
fescue (Decker et al., 1980) or 'Pensacola' bahaigrass
(Bertand et al., 1981) and sludge content of feces of
cattle which rotationally graze these pastures.

Study and Treatment	Sludge Solids	Appli-cation Rate	Sludge in/on Forage	Sludge in Feces
	%	cm	%	%
Decker et al. (1980)[*]				
1976 - 21-day sludge	4.4	20 x 0.51	5.39	7.1
1976 - 1-day sludge	4.8	20 x 0.51	22.3	18.6
1977 - 21-day sludge	2.9	20 x 0.51	2.18	7.7
1977 - compost			(0.74)[†]	6.5
1978 - 21-day sludge	3.7	24 x 0.51	2.91	6.1
1978 - compost			(0.50)[†]	2.0
Bertrand et al. (1981)[‡]				
1979 - 7-26-day sludge	2.1	9 x 0.84	2.17	4.6
1979 - 7-13-day sludge	2.1	18 x 0.84	5.17	5.8

[*]Four paddocks grazed on a rotation system; sludge was applied to
clipped pasture 21 days before grazing (21-day sludge), or regrown
pastures 1 day before grazing (1-day sludge). Compost applied 3
times in 1977 and 1 time in 1978, with at least 21 days before
grazing began.

[†]Estimates based on individual elements were not in close agreement;
no significant sludge content

[‡]Data of Bertrand et al. (1981) recalculated using results for Cu, Fe,
Pb, and Zn, elements substantially increased by sludge application.
Two paddocks were grazed in rotation. Rotations were made every
12-14 days; depending on forage growth and weather, the sludge
application occurred 7 to 13 days before grazing commenced. The
two sludge treatments differed in number of sludge applications
made during the grazing season.

plow layer soil prevents sludge adherence to forage crops, and
greatly reduces potential ingestion of sludge from the soil
surface.
   Ingestion of sludge from the soil surface was estimated for
a compost amendment which did not adhere to the crop (Decker et
al., 1980). Cattle consumed about 1-3% compost when the forage
had no detectable compost adhering (Table 39). Lower compost
ingestion occurred in 1978 when compost was applied only once
during the growing season. Others have evaluated soil consumption by well managed dairy cattle, sheep and swine (Fries et al.
1982, Fries and Marrow, 1982). Up to 8% soil was ingested from
pasture, and less from bare soil. Hogue et al. (1984) reported

metal residues in sheep tissues after the sheep grazed 152 days on a grass-legume pasture established on soil in which 224 mt/ha metal-rich sludge had been incorporated. Although the forage was increased in Cd, kidney Cd of sheep was not increased. Other element residues were not influenced by sludge incorporation. Similar results were obtained when sheep grazed pastures which received surface-applied high Cu swine manure during the previous grazing season. Neither liver Cu nor fecal Cu was consistently affected by previous manure application (Poole et al., 1983).

Ingestion of sludge-borne trace elements does not necessarily cause the health effects which are expected based on traditional toxicological studies with added metal salts. Sludge feeding studies have been conducted to evaluate element deposition in tissues of cattle, sheep, and swine. Low metal concentration sludges have not increased Cd, Zn, Pb, Zn, etc. in animal tissues in several studies (Decker et al., 1980; Baxter et al., 1982; Evans et al., 1979); while high Cd sludges have increased Cd in liver and kidney (Bertrand et al., 1980; Fitzgerald, 1980; Johnson et al., 1981; Kienholz et al., 1979; Baxter et al., 1982; Hansen et al., 1981) (see review in Hansen and Chaney, 1983). The most consistent potential problem resulting from sludge ingestion is reduced Cu concentration in the liver. Sludge Zn, Cd, Fe, and possibly Mo could interfere with Cu absorption. Ingestion of sludge rich in Fe induced Cu-deficiency in cattle in the only sludge feeding or sludge grazing study where animal performance or health was negatively affected (Decker et al., 1980). When ingested sludges are rich in Cd, Hg, F, or Pb, deposition occurs in bone or liver, but little change has been found in livestock tissues used as food (Hansen and Chaney, 1983).

Crops grown on sludge-amended soils can transfer trace elements to feeds and foods. However, the extent of increases of trace elements in crop tissues, and the bioavailability of these to animals varies with sludge properties. Crops grown on soils amended with low metal sludges had little effect on kidney Cd in several studies; however, high Cd sludges increased Cd in crops, which increased Cd in kidney and liver (Decker et al., 1980; Bertrand et al., 1980; Rundle et al., 1984; Miller and Boswell, 1979; Chaney et al. 1978a, 1978b; Boyd et al, 1982; Bray et al., 1985; Babish et al., 1979; Haschek et al., 1979; Heffron et al., 1980; Lisk et al., 1982; Ray et al., 1982; Sanson et al., 1984; Smith et al., 1985; Telford et al., 1982, 1984; Williams et al., 1978).

Much of this work has focused on Cd, because Cd can be mobile in food chains. Humans and laboratory animals have been used to characterize Cd bioavailability. Problems have been identified with the experimental methods used in this research. The early studies (Rahola et al., 1973; Yamagata et al., 1975) measured retention after only a few days or weeks. McLellan et al. (1978) found that part of the diet Cd was absorbed by intestinal mucosal cells which were subsequently sloughed into the intestine and the diet Cd repeatedly recycled in intestinal cells for a prolonged time. This delayed excretion allowed true absorption long after the test diet was fed and other parts of the test meal excreted.

Flanagan et al. (1978) found that Fe deficiency very strongly
affected Cd retention. The Fe deficiency increased Cd absorption
into the intestinal mucosal cells where it was largely trapped as
Cd-metallothionein. Fox et al. (1984) showed this aspect of Fe
deficiency allows increase in true Cd absorption and movement to
kidney long after the test diet residue is excreted. Shaikh and
Smith (1980) were able to study subjects up to 800 days after the
test dose (using [109]Cd), and resolved whole body Cd into 3 Cd
pools, now including the slowly excreted intestinal Cd turnover
pool. The biological half-life of the slowest pool was 18 years
to infinity rather than 100 days as previously reported for the
shorter-term human studies. Again, Fe deficiency affected
retention of dietary Cd, but the apparent retention was appreciably lower than in earlier studies. These tend to support the
findings of Newton et al. (1984) and Sharma et al. (1983) that
only low amounts of Cd are retained by humans ingesting Western-
type diets.

CONCLUSIONS

1. Conditions for valid assessment of relative increased crop
    concentration of an element due to sludge utilization are
    limited to long-term sludge amended soils, preferably 2 or
    more years after sludge is applied. Metal salts and metal
    salt-amended sludges do not provide valid data for assessment
    of food-chain element transfer.

2. Some trace elements can be increased in edible crop tissues
    when sewage sludges rich in the element are applied to acidic
    soils (Cd, Zn, Ni), or alkaline soil (Mo). Under these con-
    ditions which allow substantial increase of a trace element
    in crops (responsive conditions), the relative increase in
    element concentration among crop species are sufficiently
    consistent to be relied upon in dietary exposure modeling.
    Some variation in relative increased trace element con-
    centration among crops may result from high soil organic
    matter, or from calcareous vs. acidic soil conditions.
    High organic matter and high soil pH both reduce element
    uptake (except for Mo and Se).

3. Except for corn inbreds, cultivar variation in element con-
    centration has been found to be approximately 2- to 5-fold.
    Because of inclusion of various cultivars in the food supply,
    this variation would not significantly alter chronic exposure
    due to increased crop uptake of sludge applied elements.
    However, cultivar selection can be used to reduce food-chain
    transfer of elements from well managed sludge utilization
    farms.

4. If the FDA food groups are used in dietary Cd modeling, they
    should be adjusted for relative high and low Cd accumulating
    crop types (lettuce vs. cabbage; carrot vs. beet) within a
    food group. Food intakes should represent average adult

intake for 50 years, not the maximum intakes of teen-age
males. For evaluation of potential chronic Cd exposure from
acidic sludge-amended garden soils, the adult food intakes
reported by Pennington (1983) can be used, and the relative
increases in Cd in crops or food groups summarized in this
report. Increased Cd uptake by all garden foods can be integrated in terms of increased Cd uptake by a reference crop
such as lettuce. Dietary Cd increase can be predicted by the
response of Cd concentration in lettuce grown in test soils
(height of the plateau in lettuce Cd on sludge-amended soils
above the control, as affected by sludge Cd concentration and
other factors) times the integrated garden foods Cd intake
factor. Thus, increase in dietary Cd due to growing 100% of
consumed garden vegetables on sludge-amended acidic garden
soils was estimated as 2.20 µg Cd/d when lettuce is increased
above background by 1 mg/kg dry lettuce.

5. Prediction of changes in kidney Cd due to increases in
   dietary Cd from foods grown in acidic sludge-amended gardens
   should consider effects of nutritional status and nutrients
   in the garden crops on Cd retention by humans.

6. Ingestion of sludge can allow exposure and/or risk which can
   be prevented by incorporation of sludge below the soil surface, or by tilling sludge into the soil. For some elements
   (e.g., Fe, F, Cu, Zn, Pb), direct sludge ingestion may allow
   sufficient exposure to sludge-borne elements to cause risk,
   at least for element-rich sludges. Bioavailability of many
   elements in ingested sludge is very strongly influenced by
   concentration of the element and other elements present,
   sludge carbonate content, and sludge redox potential. For
   many elements which comprise potential risk if sludge is
   ingested, median quality sludges have not caused any problems
   with livestock at common exposure rates from surface-applied
   sludge.

REFERENCES

Adams, C. F. 1975. Nutritive values of American foods in common
units. Agricultural handbook No. 456, U.S. Government Printing
Office, 0100-03184.

Babish, J. G., G. S. Stoewsand, A. K. Furr, T. F. Parkinson,
C. A. Bache, W. H. Gutenmann, P. C. Wszolek, and D. J. Lisk.
1979. Elemental and polychlorinated biphenyl content of tissues
and intestinal aryl hydrocarbon hydroxylase activity of guinea
pigs fed cabbage grown on municipal sewage sludge. J. Agr.
Food Chem. 27:399-402.

Bache, C. A. W. H. Gutenmann, W. D. Youngs, J. G. Doss and
D. J. Lisk. 1981. Absorption of heavy metals from sludge-
amended soil by corn cultivars. Nutr. Rep. Int. 23:499-503.

Baxter, J. C., B. Barry, D. E. Johnson and E. W. Kienholz. 1982. Heavy metal retention in cattle tissues from ingestion of sewage sludge. J. Environ. Qual. 11:616-620.

Beaton, G. H., J. Milner, V. McGuire, T. E. Feabher and J. A. Little. 1983. Source of variance in 24-hour dietary recall data: Implications for nutrition study design and interpretation. Carbohydrate sources, vitamins and minerals. Am. J. Clin. Nutr. 37:986-995.

Bertrand, J. E., M. C. Lutrick, H. L. Breland, and R. L. West. 1980. Effects of dried digested sludge and corn grown on soil treated with liquid digested sludge on performance, carcass quality, and tissue residues in beef cattle. J. Anim. Sci. 50:35-40.

Bertrand, J. E., M. C. Lutrick, G. T. Edds, and R. L. West. 1981. Metal residues in tissues, animal performance and carcass quality with beef steers grazing Pensacola bahiagrass pastures treated with liquid digested sludge. J. Anim. Sci. 53:146-153.

Black, G. 1982. A review of validations of dietary assessment methods. Am. J. Epidem. 115:492-505.

Boggess, S. F., S. Willavize and D. E. Koeppe. 1978. Differential of soybean varieties to soil cadmium. Agron. J. 70:756-760.

Boyd, J. N, G. S. Stoewsand, J. G. Babish, J. N. Telford and D. J. Lisk. 1982. Safety evaluation of vegetables cultured on municipal sewage sludge-amended soil. Arch. Environ. Contam. Toxicol. 11:399-405.

Bray, B. J., R. H. Dowdy, R. D. Goodrich and D. E. Pamp. 1985. Trace metal accumulations in tissues of goats fed silage produced on sewage sludge-amended soil. J. Environ. Qual. 14:114-118.

Burau, R. G. 1980. Current knowledge of cadmium in soils and plants as related to cadmium levels in foods. pp. 65-72. In: Proc. 1980 TFI Cadmium Seminar, The Fertilizer Institute, Washington, D.C.

Carlton-Smith, C. H. and R. D. Davis. 1983. Comparative uptake of heavy metals by forage crops grown on sludge-treated soils. pp. 393-396. In: Proc. Int. Conf. Heavy Metals in the Environment. Vol. 1, Heidelberg, 1983. CEP Consultants, Ltd., Edingburgh, UK.

Council for Agricultural and Scientific Technology. 1980. Effects of sewage sludge on the cadmium and zinc content of crops. Council for Agricultural Science and Technology Report No. 83. Ames, IA. 7 pp.

Chaney, R. L. 1985. Personal communication, USDA-ARS-NER, U.S. Environmental Protection Agency, Cincinnati, OH 45268.

Chaney, R. L. and. C. A. Lloyd. 1979. Adherence of spray-applied liquid digested sewage sludge to tall fescue. J. Environ. Qual. 8:407-411.

Chaney, R. L. and C. A. Lloyd. 1986. Personal communication. USDA-ARS-NER, U.S. Environmental Protection Agency, Cincinnati, OH 45268.

Chaney, R. L., G. S. Stoewsand, C. A. Bache and D. J. Lisk. 1978a. Cadmium deposition and hepatic microsomal induction in mice fed lettuce grown on municipal sludge-amended soil. J. Agric. Food Chem. 26:992-994.

Chaney, R. L., G. S. Stoewsand, A. K. Furr, C. A. Bache and D. J. Lisk. 1978b. Elemental content of tissues of guinea pigs fed Swiss chard grown on municipal sewage sludge-amended soil. J. Agr. Food Chem. 26:994-997.

Chang, A. C., A. L. Page, K. W. Foster and T. E. Jones. 1982. A comparison of cadmium and zinc accumulation by four cultivars of barley grown in sludge-amended soils. J. Environ. Qual. 11:409-412.

Chang, A. C., A. L. Page, L. J. Lund, P. F. Pratt, and G. R. Bradford. 1978. Land application of sewage sludge--A field demonstration. Regional Wastewater Solids Management Program, Los Angeles/Orange County Metropolitan Area (LAOMA), Dept. of Soil and Environmental Sciences, Univ. of California, Riverside, CA.

Chino, M. 1981. Uptake-transport of toxic metals in rice plants. pp. 81-94. In: (K. Kitagishi and I. Yamane, eds.) Heavy metal pollution of soils of Japan. Japan Sci. Soc. Press, Tokyo.

Chino, M. and A. Baba. 1981. The effects of some environmental factors on the partitioning of zinc and cadmium between roots and tops of rice plants. J. Plant Nutr. 3:203-214.

Crews, H. M., and B. E. Davies. 1985. Heavy metal uptake from contaminated soils by six varieties of lettuce (Lactuca sativa L.). J. Agric. Sci. 105:591-595.

Davies, B. F., and N. J. Lewis. 1985. Controlling heavy metal uptake through choice of plant varieties. p. 102-108. In: Proc. 19th Conf. on Trace Substances in Environmental Health, Univ. of Missouri, Columbia, MO.

Davis, R. D. and C. Carlton-Smith. 1980. Crops as indicators of the significance of contamination of soil by heavy metals. 44 pp. Technical Rpt. TR140, Water Research Centre, Stevenage, UK.

Dean, R. B. and M. J. Suess (eds.). 1985. The risk to health of chemicals in sewage sludge applied to land. Waste Manag. Res. 3:251-278.

Decker, A. M., R. L. Chaney, J. P. Davidson, T. S. Rumsey, S. B. Mohanty, and R. C. Hammond. 1980. Animal performance on pastures top-dressed with liquid sewage sludge and sludge compost. pp. 37-41. In: Proc. Nat. Conf. Municipal Industrial Sludge Utilization and Disposal. Information Transfer, Inc., Silver Spring, MD.

Dowdy, R. H., and W. E. Larson. 1975. The availability of sludge-borne metals to various vegetable crops. J. Environ. Qual. 4:278-282.

Environmental Protection Agency. 1979a. Criteria for classification of solid waste disposal facilities and practices; interim final, and proposed regulations. 40 CFR 257. Docket No. 4004. Fed. Reg. 44(179):53438-53468.

Environmental Protection Agency. 1979b. Background document: Cumulative cadmium application rates. Criteria for classification of solid wastes disposal facilities. 40 CFR 257. Sept. 1979. US EPA Off. Solid Wastes.

Evans, K. J. , I. G. Mitchell and B. Salan. 1979. Heavy plant accumulation in soils irrigated by sewage and effect in the plant-animal system. Progr. Water Technol. 11:339-352.

Feder, W. A., R. L. Chaney, C. E. Hirsch and J. B. Munns. 1980. Differences in Cd and Pb accumulation among lettuce cultivars and metal pollution problems in urban soils. p. 347. In: (G. Bitton et al., eds.) Sludge—Health risks of land application. Ann Arbor Science Publishers, Ann Arbor, MI.

Fitzgerald, P. R. 1980. Observations on the health of some animals exposed to anaerobically digested sludge originating in the Metropolitan Sanitary District of Greater Chicago system. pp. 267-284. In G. Bitton et al. (eds.). Sludge—Health Risks of Land Application. Ann Arbor Sci. Publ. Inc.

Flanagan, P. R., J. S. McLelland, J. Haist, G. Cherian, M. J. Chamgerlain and L. S. Walberg. 1978. Increased dietary cadmium absorption in mice and human subjects with iron deficiency. Gastroenterol. 74:841-846.

Flynn, Michael P. 1981. Personal communication, Office of Solid Waste, US EPA, Washington, D.C.

Flynn, Michael P. 1986. Personal communication, Office of Solid Waste, US EPA, Washington, D.C.

Fox, M. R. S., S. H. Tao, C. L. Stone and B. E. Fry, Jr. 1984. Effects of zinc, iron and copper deficiencies on cadmium in tissues of Japanese quail. Environ. Health Perspect. 54:57-65.

Foy, C. D., R. L. Chaney an M. C. White. 1978. The physiology of dairy cattle. Ann. Rev. Plant Physiol. 29:511-566.

Francis, C. W., E. C. Davis, and J. C. Goyert. 1985. Plant uptake of trace elements from coal gasificiation ashes. J. Environ. Qual. 14:561-569.

Fries, G. F. 1982. Potential polychlorinated biphenyl residues in animal products from application of contaminated sewage sludge to land. J. Environ. Qual. 11:14-20.

Fries, G. F., and G. S. Marrow. 1982. Soil ingestion by swine as a route of contaminant exposure. Environ. Toxicol. Chem. 1:201-204.

Fries, G. F., G. S. Marrow and P. A. Snow. 1982. Soil ingestion by dairy cattle. J. Dairy Sci. 65:611-618.

Furr, A. K., A. W. Lawrence, S. S. C. Tong, M. C. Grandolfo, R. A. Hofstader, C. A. Bache, W. H. Gutenmann, and D. J. Lisk. 1976a. Multielement and chlorinated hydrocarbon analysis of municipal sewage sludges of American cities. Environ. Sci. Technol. 10:683-687.

Furr, A. K., W. C. Kelly, C. A. Bache, W. H. Gutenmann, and D. J. Lisk. 1976b. Multielement absorption by crops grown on Ithaca sludge-amended soil. Bull. Environ. Contam. Toxicol. 16:756-763.

Giordano, P. M., and D. A. Mays. 1977. Effect of land disposal applications of municipal wastes on crop yields and heavy metal uptake. Municipal Environmental Research Laboratory, Office of Research and Development, U.S. Environmental Protection Agency, Cincinnati, OH.

Giordano, P. M., D. A. Mays and A. D. Behel, Jr. 1979. Soil temperature effects on uptake of cadmium and zinc by vegetables grown on sludge-treated soil. J. Environ. Qual. 8:233-236.

Grill, E., E. L. Winnacker, and M. H. Zenk. 1985. Phytochelatins: The principal heavy-metal complexing peptides of higher plants. Science 230:674-676.

Hallberg, L., E. Bjorn-Rasmussen, L. Rossander and R. Suwanik. 1977. Iron absorption from southeast Asian diets. II. Role of various factors that might explain low absorption. Am. J. Clin. Nutr. 30:539-548.

Hallberg, L., L. Garby, R. Suwanik and E. Bjorn-Rasmussen. 1974. Iron absorption from southeast Asian diets. Am. J. Clin. Nutr. 27:826-836.

Hansen, L. G., P. K. Washko, L. G. M. T. Tuinstra, S. B. Dorn, and T. D. Hinesly. 1981. Polychlorinated biphenyl, pesticide, and heavy metal residues in swine foraging on sewage sludge amended soils. J. Agr. Food Chem. 29:1012-1017.

Hansen, L. G. and R. L. Chaney. 1983. Environmental and food chain effects of the agricultural use of sewage sludges. Rev. Environ. Toxicol. 1:103-117.

Harris, M. R., S. J. Harrison, N. J. Wilson and N. W. Lepp. 1981. Varietal differences in trace metal partitioning by six potato cultivars grown on contaminated soil. pp. 399-402. In: Proc. Int. Conf. Heavy Metals in the Environment. CEP Consultants, Ltd., Edinburgh, UK.

Harrison, H. C. 1986a. Response of lettuce cultivars to sludge amended soils and bed types. Commun. Soil Sci. Plant Anal. 17: 159-172.

Harrison, H. C. 1986b. Carrot response to sludge application and bed type. J. Am. Soc. Hort. Sci. 111:211-215.

Haschek, W. M., A. K. Furr, T. F. Parkinson, C. L. Heffron, J. T. Reid, C. A. Bache, P. C. Wszolek, W. H. Gutenmann, and D. J. Lisk. 1979. Element and polychlorinated biphenyl deposition and effects in sheep fed cabbage grown in municipal sewage sludge. Cornell Vet. 69:302-314.

Heffron, C. L., J. T. Reid, D. C. Elfving, G. S. Stoewsand, W. M. Haschek, J. N. Telford, A. K. Furr, T. F. Parkinson, C. A. Bache, W. H. Gutenmann, P. C. Wszolek, and D. J. Lisk. 1980. Cadmium and zinc in growing sheep fed silage corn grown on municipal sludge amended soil. J. Agr. Food Chem. 28:58-61.

Hemphill, D. D., Jr., T. L. Jacobson, L. W. Martin, G. L. Kiemnec, D. Hanson, and V. V. Volk. 1982. Sweet corn response to application of three sewage sludges. J. Environ. Qual. 11:191-196.

Hinesly, T. D. 1986. Personal communication, Dept. of Agronomy, University of Illinois, Urbana, IL 61801.

Hinesly, T. D., D. E. Alexander, K. E. Redborg and E. L. Ziegler. 1982a. Differential accumulations of cadmium and zinc by corn hybrids grown on soil amended with sewage sludge. Agron. J. 74:469-474.

Hinesly, T. D., K. E. Redborg, E. L. Ziegler, and J. D. Alexander. 1982b. Effect of soil cation exchange capacity on the uptake of cadmium by corn. Soil Sci. Soc. Am. J. 46:490-497.

Hinesly, T. D., D. E. Alexander, E. L. Ziegler and G. L. Barrett. 1978. Zinc and Cd accumulation by corn inbreds grown on sludge amended soil. Agron. J. 70:425-428.

Hinesly, T. D., L. G. Hansen, D. J. Bray and K. E. Redborg. 1985. Transfer of sludge-borne cadmium through plants to chickens. J. Agr. Food Chem. 33:173-180.

Hinesley, T. D., V. Sudarski-Hack, D. E. Alexander, E. L. Ziegler, and G. L. Barrett. 1979. Effect of sewage sludge applications on phosphorus and metal concentrations in fractions of corn and wheat kernels. Cereal Chem. 56:283-287.

Hogue, D. E., J. J. Parrish, R. H. Foote, J. R. Stouffer, J. L. Anderson, G. S. Stoewsand, J. N. Telford, C. A. Bache, W. H. Gutenmann, and D. J. Lisk. 1984. Toxicologic studies with male sheep grazing on municipal sludge-amended soil. J. Toxicol. Environ. Health 14:153-161.

Jarvis, S. C., L. H. P. Jones, and M. J. Hopper. 1976. Cadmium uptake from solution by plants and its transport from roots to shoots. Plant Soil 44:179-191.

Johnson, D. E., E. W. Kienholz, J. C. Baxter, E. Spanger and G. M. Ward. 1981. Heavy metal retention in tissues of cattle fed high cadmium sewage sludge. J. Anim. Sci. 52:108-114.

Jones, S. G., K. W. Brown, L. E. Deuel and K. C. Donnelly. 1979. Influence of simulated rainfall on the retention of sludge heavy metals by the leaves of forage crops. J. Environ. Qual. 8:69-72.

Kienholz, E. W., G. M. Ward, D. E. Johnson, J. Baxter, G. Braude, and G. Stern. 1979. Metropolitan Denver sewage sludge fed to feedlot steers. J. Anim. Sci. 48:735-741.

Korcak, R. F. 1986. Personal communication, USDA-ARS, Beltsville, MD 20705.

Korcak, R. F., and D. S. Fanning. 1985. Availability of applied heavy metals as a function of type of soil material and metal source. Soil Sci. 140:23-34.

Lisk, D. J., R. D. Boyd, J. N. Telford, J. G. Babish, G. S. Stoewsand, C. A. Bache, and W. H. Gutenmann. 1982. Toxicologic studies with swine fed corn grown on municipal sewage sludge-amended soil. J. Anim. Sci. 55:613-619.

Logan, T. J. and R. L. Chaney. 1983. Utilization of municipal wastewater and sludge on land--Metals. p. 235-326. In A. L. Page et al. (ed.) Proc. of the workshop on utilization of municipal wastewater and sludge on land. University of California, Riverside, CA.

McKenzie, J., T. Kjellstrom, and R. Sharma. 1982. Cadmium intake, metabolism and effects in people with a high intake of oysters in New Zealand. Draft Project Report to EPA. Grant #R807058-01-0.

McLellan, J. S., P. R. Flanagan, M. J. Chamberlain, and L. S. Valberg. 1978. Measurement of dietary cadmium absorption in humans. J. Toxicol. Environ. Health 4:131-138.

Meyer, M. W., F. L. Fricke, G. G. S. Holmgren, J. Kubota, and
R. L. Chaney. 1982. Cadmium and lead in wheat grain and
associated surface soils in major wheat production areas of
the United States. Agron. Abst. American Society of Agronomy,
Madison, WI. p. 34.

Miller, J., and F. C. Boswell. 1979. Mineral content of
selected tissues and feces of rats fed turnip greens grown
on soil treated with sewage sludge. J. Agr. Food Chem.
27:1361-1365.

Newton, D., P. Johnson, A. E. Lally, R. J. Pentreath, and
D. J. Swift. 1984. The uptake by man of cadmium ingested in
crab meat. Human Toxicol. 3:23-28.

Pedersen, B. and B. O. Eggum. 1983. The influence of milling
on the nutritive value of flour from cereal grains. 4. Rice.
Qual. Plant. Plant Foods Hum. Nutr. 33:267-278.

Pennington, J. A. T. 1983. Revision of total diet study food
lists and diets. J. Am. Diet. Assoc. 82:166-173.

Pepper, I. L., D. E. Bezdicek, A. S. Baker, and J. M. Sims.
1983. Silage corn uptake of sludge-applied zinc and cadmium as
affected by soil pH. J. Environ. Qual. 12:270-275.

Poole, D. B. R., D. McGrath, G. A. Fleming, and J. Sinnott.
1983. Effects of applying copper-rich pig slurry to grassland.
3. Grazing trials: Stocking rate and slurry treatment. Ir. J.
Agric. Res. 22:1-10.

Rahola, T., R.-K. Aaran and J. K. Miettinen. 1973. Retention
and elimination of [115m]Cd in man. pp. 213-218. In: (E.
Bujdosa, ed.) Health physics problems of internal contamination.
Akademiai Kiado, Budapest.

Rauser, W. E., and J. Glover. 1984. Cadmium-binding protein
in roots of maize. Can. J. Bot. 62:1645-1650.

Ray, E. E., R. T. O'Brien, D. M. Stiffler, and G. S. Mith. 1982.
Meat quality from steers fed sewage solids. J. Food Protec. 45:
317-323.

Robinson, N. J. and P. J. Jackson. 1986. 'Metallothionein-like'
metal complexes in angiosperms; their structure and function.
Physiol. Plant. 67:499-506.

Romheld, V., and H. Marschner. 1986. Evidence for a specific
uptake system for iron phytosiderophores in roots of grasses.
Plant Physiol. 80:175-180.

Rundle, H. L., M. Calcroft and C. Holt. 1984. An assessment of
accumulation of Cd, Cr, Cu, Ni and Zn in the tissues of British
Friesian steers fed on the products of land which has received
heavy applications of sewage sludge. J. Agr. Sci. 102:1-6.

Ryan, J. A., H. R. Pahren and J. B. Lucas. 1982. Controlling cadmium in the human food chain: A review and rationale based on health effects. Environ. Res. 28:251-302.

Sanson, D. W., D. M. Hallford, and G. S. Smith. 1984. Effects of long-term consumption of sewage solids on blood, milk, and tissue elemental composition of breeding ewes. J. Anim. Sci. 59:416-424.

Sempos, C. T., N. E. Johnson, E. L. Smith, and C. Gilligan. 1985. Effects of intraindividual and interindividual variation in repeated dietary records. Am. J. Epidemiol. 121:120-130.

Shaikh, Z. A. and J. C. Smith. 1980. Metabolism of orally ingested cadmium in humans. pp. 569-574. In: (B. Holmstedt et al. eds.) Mechanisms of toxicity and hazard evaluation. Elsevier/North-Holland Biomedical Press.

Sharma, R. P., T. Kjellstrom and J. M. McKenzie. 1983. Cadmium in blood and urine among smokers and non-smokers with high cadmium intake via food. Toxicol. 29:163-171.

Sharma, R. P., J C. Street, M. P. Verma and J. L. Shupe. 1979. Cadmium uptake from feed and its distribution to food products of livestock. Environ. Health Perspect. 28:59-66.

Smith, G. S., D. M. Hallford, and J. B. Watkins, III. 1985. Toxicological effects of gamma-irradiated sewage solids fed as seven percent of diet to sheep for four years. J. Anim. Sci. 61:931-941.

Sugiura, Y., and K. Nomoto. 1984. Phytosiderophores. Structures and properties of mugineic acids and their metal complexes. Structure and Bonding 58:106-135.

Telford, J. N., J. G. Babish, B. E. Johnson, M. L. Thonney, W. B. Currie, C. A. Bache, W. H. Gutenmann, and D. J. Lisk. 1984. Toxicologic studies with pregnant goats fed grass-legume silage grown on municipal sludge-amended subsoil. Arch. Environ. Contam. Toxicol. 13:635-640.

Telford, J. N., M. L. Thonney, D. E. Hogue, J. R. Stouffer, C. A. Bache, W. H. Guttenmann, D. J. Lisk, J. G. Babish, and G. S. Stoewsand. 1982. Toxicologic studies in growing sheep fed silage corn cultured on municipal sludge-amended acid subsoil. J. Toxicol. Environ. Health 10:73-85.

Todd, K. S., M. Hudes and D. H. Calloway. 1983. Food intake measurement: Problems and approaches. Am. J. Clin. Nutr. 37:139-146.

Tsuchiya, K. (ed.) 1978. Cadmium studies in Japan: A review. Elsevier/North-Holland Biomedical Press, New York. 376 pp.

Williams, P. H., J. S. Shenk, and D. E. Baker. 1978. Cadmium accumulation by meadow voles (Microtus pennsylvanicus) from crops grown on sludge-treated soil. J. Environ. Qual. 7:450-454.

Wolnik, K. A., F. L. Fricke, S. G. Capar, G. C. Braude, M. W. Meyer, R. D. Satzger, and E. Bonnin. 1983a. Elements in major raw agricultural crops in the United States. 1. Cadmium and lead in lettuce, peanuts, potatoes, soybeans, sweet corn, and wheat. J. Agric. Food Chem. 31:1240-1244.

Wolnik, K. A., F. L. Fricke, S. G. Capar, G. C. Braude, M. W. Meyer, R. D. Satzger, and R. W. Kuennen. 1983b. Elements in major raw agricultural crops in the United States. 2. Other elements in lettuce, peanuts, potatoes, soybeans, sweet corn and wheat. J. Agric. Food Chem. 31:1244-1249.

Wolnik, K. A., F. L. Fricke, S. G. Capar, M. W. Meyer, R. D. Satzger, E. Bonnin, and C. M. Gaston. 1985. Elements in major raw agricultural crops in the United States. 3. Cadmium, lead, and eleven other elements in carrots, field corn, onions, rice, spinach, and tomatoes. J. Agric. Food Chem. 33:807-811.

Yamagata, N., K. Iwashima and T. Nagai. 1975. Gastrointestinal absorption of ingested ^{115m}Cd by a man. Bull. Inst. Public Health 24:1-6.

Yoshikawa, T., S. Kusaka, T. Zikihara and T. Yoshida. 1977. Distribution of heavy metals in rice plants. I. Distribution of heavy metal elements in rice grains using an electron probe x-ray microanalyzer (EPMA). J. Sci. Soil Manure, Japan 48:523-528. (In Japanese).

Yuran, G. T. and H. C. Harrison. 1986. Effects of genotype and sewage sludge on cadmium concentration in lettuce leaf tissue. J. Am. Soc. Hort. Sci. 111:491-494.

CHAPTER 6

EFFECTS OF TRACE ORGANICS IN SEWAGE SLUDGES ON
SOIL-PLANT SYSTEMS AND ASSESSING THEIR RISK TO HUMANS

Lee W. Jacobs
Michigan State University, East Lansing, Michigan

George A. O'Connor
New Mexico State University, Las Cruces, New Mexico

Michael A. Overcash
North Carolina State University, Raleigh, North Carolina

Matthew J. Zabik
Michigan State University, East Lansing, Michigan

Paul Rygiewicz
U.S. Environmental Protection Agency, Corvallis, Oregon

In consultation with:
Peter Machno and Sydney Munger
Metropolitan Seattle Water Quality Division, Seattle, Washington

Ahmed A. Elseewi
Southern California Edison Company, Rosemead, California

INTRODUCTION

Describing the impact of trace organics in sludge on soil-
plant systems can be an even greater challenge than is faced
with trace elements. One reason is the sheer number of compounds
potentially involved. Literally thousands of trace organics
exist and many, if not all, can be expected in sewage sludge at
highly variable concentrations. At the same time, the literature
discussing the effects of trace organics on soil-plant systems is
much less voluminous than the trace element literature. Undoubtedly, the paucity of scientific studies on trace organics is due
to the complexity of studying these chemicals and the expense of
trace organic analyses.

An important difference between trace metal and trace organic
additions to a soil is the time each may reside or persist in

that soil. The half-life of the most persistent organics (e.g., PCBs) in soil was concluded to be 10 or more years (Fries, 1982), whereas the residence time for most metals was estimated to be a few thousand years (Bowen, 1977). Studies of trace organic behavior in soils must also consider assimilation mechanisms such as degradation (biotic and abiotic) and volatilization, in addition to factors such as solubility, adsorption/desorption, leaching and plant uptake. While these additional mechanisms make trace organic studies more challenging, they also lend themselves to management alternatives not available for trace elements. For example, long-term application programs with organics (e.g., food processing and petroleum wastes) attest to the soil's ability to receive and successfully assimilate wastes over time.

This attenuation capacity suggests that limiting the addition of trace organics to soils via sludge application should be based on "matching the total loadings of an organic(s) with the soil's assimilative ability" rather than "a specified concentration present in the sludge". Such an approach allows more flexibility to consider environmentally sound options on a case by case basis for local circumstances.

Some analogies of trace organics to agricultural pesticides can be made. But the influence of the sludge organic matter matrix in combination with a specific organic, when added to the soil, is poorly understood.

To facilitate discussion, the large number of organic chemicals were divided into groups which tend to have similar chemical and physical properties. Various organics could then be discussed by groups relevant to their prevalence in sludge, fates in the soil-plant system, and recent efforts at assessing the risk of trace organic additions to the soil via sludge application.

The following discussion focuses on the impact of trace organics to soil-plant systems. The toxicity of trace organics to soil organisms, animals or humans as a result of their addition to soils and potential pathways, whereby exposure of soil-applied organics to animals and humans might occur, are listed and briefly discussed.

PREVALENCE OF TRACE ORGANICS IN SLUDGES

Any program to assess the risk from trace organics must begin by determining (1) which chemicals are the most likely to be present in sewage sludge and (2) what quantities may be added to soils by the application of sludges containing these trace organics. To assess the potential impacts, priority should be given to substances shown to be prevalent in sewage sludge through residue analysis and which have certain physical-chemical properties that could lead to unacceptable toxicological or environmental effects. The next priority are those chemicals heavily used in society, possessing similar undesirable physical-chemical properties but not yet identified in sludges. With the current data base only the first priority can be effectively considered.

Since municipal raw sewage contains virtually all the wastes from man's activities, one could expect the sludge resulting from the treatment of this sewage to also contain these same products. Because as many as 15,000-20,000 man-made chemicals (with an array of functional groups) exist, analyzing all the chemical constituents in a sewage sludge is impossible. Sludges have been analyzed according to predetermined lists of specific organic chemicals such as the organic priority pollutants list (NRDC, 1976). Another approach is to separate the organics into "chemical groups" which have similar physical-chemical properties and focus on selected groups anticipated to have a greater toxicological and/or environmental risk. With either approach, the wastewater treatment plant should first determine what organics are being discharged by users, particularly industry. This information can then be used as guidance for identifying which organic compounds should be tested to check for any unusually high concentrations in sewage sludges.

A third approach is to use short-term bioassays to test sludges or sludge-amended soil for mutagenicity (Brown et al., 1982; Hopke and Plewa, 1984; and Peters, 1985). Sludges failing such a test would then be evaluated more rigorously and analyzed for selected organics. This approach requires additional research on suitable bioassays followed by calibration of these bioassays with experiences in the field, before it could realistically be used.

Only a few studies have reported the analyses of trace organics in sewage sludges. These studies confirm the wide variety of trace organic compounds that can occur in sewage sludge, but significant problems exist in the analysis and interpretation of these data:

- Sludges are heterogeneous and obtaining a representative sample can be difficult.

- Day-to-day variations in composition occur.

- Analytical protocols vary widely in extractions, separations and cleanup procedures which in turn affect the number and types of compounds recovered.

- For some groups, recoveries from a complex matrix like sludge can be poor.

- Data are reported in various units ($\mu g/l$, mg/kg, etc., some on a wet weight basis, and others on dry weight).

- Limits of detection in some cases are poor or are not reported.

- Confirmation of each organic, if any was done, is not reported.

Because of these problems and a very limited data base, definitive statements concerning the prevalence of organic

chemicals in sewage sludge can not be made. To rectify this
situation, the following information and data are suggested as a
minimum for reporting on the organic content of sewage sludge:
(a) type of sludge, (b) percent dry solids, (c) number of samples
analyzed, (d) number of "positive" samples above detection limit,
and the following based on dry weight —maximum and minimum concentration (range), (e) detection limit, and (f) the median concentration of all samples tested. This information would provide
a means of standardization for comparing data sets.

Residual levels of trace organic compounds found in sewage
sludge analysis surveys are listed in Table 40. The majority of
these data come from two sources (Burns and Roe, 1982; and Jacobs
and Zabik, 1983) but are supplemented by several others. Studies
reporting organic concentrations for fewer than 9 sludges were
not included in this summary, except for one which provided data
for dioxins and furans (Weerasinghe et al., 1985). Some limitations of the data reviewed (Table 40) are that detection limits
were not reported, some data were reported on a wet weight basis
without "% solids" values given, and median concentrations for
all samples were not provided.

Compounds were listed under the following major groups based
on similar physical-chemical characteristics:

- phthalate esters
- monocyclic aromatics
- polynuclear aromatics (PAH's)
- halogenated biphenyls (PCB's)
- dioxins and furans
- halogenated aliphatics
  (short chain)

- triaryl phosphate esters
- aromatic and alkyl amines
- phenols
- chlorinated pesticides
  and hydrocarbons
- miscellaneous compounds

These data show that sewage sludges can be highly contaminated with organic chemicals. Unusually high concentrations,
such as the maximum levels shown for butylbenzylphthalate,
bis(2-ethylhexyl) phthalate, toluene, methyl bromide, chloro-
ethane, vinyl chloride, pentachlorophenol and others, suggest
a high degree of industrial contamination. Sludges containing
these "maximum concentrations" could have a significant impact
on soil-plant systems, depending on the rate of sludge applied.

Concerns about organic chemicals in sludges must be kept in
perspective, however. Of the 219 organic chemicals collectively
measured in sludges, 70 (or 32%) were below detection limits
(Table 41). About one-fourth (53) of these organics were present
in >50% of the sludges (Table 41). The presence of "background
concentrations" of many organics in purely domestic sewage
sludges is not unexpected, given the wide variety of synthetic
organic chemicals found in many household products (Hathaway,
1980). The fact that domestic septic tank effluents contain
greater than 100 trace level organics provides additional evidence for their presence in household wastewaters (DeWalle et
al., 1985; Tomson et al., 1984).

More important than the presence of an organic(s) in sewage
sludge is the total amount which may get applied to the soil-

Table 40. Summary of organic chemicals found in sewage sludges.

Chemical	No. of sludges tested	% Occur-rences	Concentrations for samples testing positive, i.e., > detection limits mg/kg (dry wt.)$ Range	Median	ug/L Range	Ref.[++]
Phthalate esters						
Bis(2-ethylhexyl)phthalate	234	84	0.415-58,300	168		1
	437	95			2-47,000	2
Butylbenzylphthalate	234	60	0.0469-12,800	59.1		1
	437	43			2-45,000	2
Diethylphthalate	234	63	0.0987-3,780	50.0		1
	437	9			1-786	2
Dimethylphthalate	236	23	0.106-941	11.7		1
	437	5			3-650	2
Di-n-butylphthalate	237	45	0.0776-3,210	17.3		1
	437	45			1-6,900	2
Di-n-octylphthalate	237	40	0.0222-2,610	4.9		1
	437	10			4-1,024	2
Monocyclic aromatics						
Benzene	436	61			1-953	2
Chlorobenzene	158	6	2.06-846	10.2		1
	436	13			1-687	2
1-chloro-2,4-dinitrobenzene	238	0	All samples < detection limit			1
1-chloro-2,6-dinitrobenzene	238	0	All samples < detection limit			1
1-chloro-3,4-dinitrobenzene	238	0	All samples < detection limit			1
1-chloro-2-nitrobenzene	238	0	All samples < detection limit			1
1-chloro-4-nitrobenzene	238	0	All samples < detection limit			1
p-chlorotoluene	158	11	1.13-324	14.7		1
1,2-dichlorobenzene	215	47	0.0229-809	0.645		1
	437	16			3-1,319	2
1,3-dichlorobenzene	215	54	0.0245-1,650	1.76		1
	437	9			14-1,900	2
1,4-dichlorobenzene	215	66	0.0402-633	2.02		1
	437	17			2-12,000	2
2,4-dinitrotoluene	238	0	All samples < detection limit			1
	431	0	All samples < detection limit			2
2,6-dinitrotoluene	238	0	All samples < detection limit			1
	431	0	All samples < detection limit			2
Ethylbenzene	220	6	1.22-65.5	19.8		1
	436	63			1-4,200	2
Hexachlorobenzene	237	43	0.000188-26.2	0.018		1
	437	2			28-780	2
Nitrobenzene	238	0	All samples < detection limit			1
	431	0	All samples < detection limit			2
Pentachloronitrobenzene	238	0	All samples < detection limit			1
Styrene	219	10	1.53-5,850	26.6		1
Toluene	434	94			1-427,000	2
1,2,3-trichlorobenzene	215	37	0.00278-152	0.0667		1
1,2,4-trichlorobenzene	217	57	0.00551-51.2	0.274		1
	437	13			2-8,300	2
1,3,5-trichlorobenzene	217	33	0.00502-39.7	0.0632		1
1,2,3,4-tetrachlorobenzene	238	0	All samples < detection limit			1
1,2,3,5-tetrachlorobenzene	238	0	All samples < detection limit			1
1,2,4,5-tetrachlorobenzene	238	0	All samples < detection limit			1
Polynuclear aromatics (PAH)						
2-chlorophthalene	437	0.2			1,600	2
Naphthalene	236	50	0.0554-6,610	30.3		1
	437	34			1-5,200	2
Acenaphthene	437	5			6-4,600	2

Table 40. (continued)

Chemical	No. of sludges tested	% Occur-rence*	Concentrations for samples testing positive, i.e., > detection limits		
			mg/kg (dry wt.)		µg/L
			Range	Median	Range	Ref.

Polynuclear aromatics (PAH) (cont'd.)

Fluorene	437	6			1-1,300	2
Phenanthrene	437	53			1-10,100	2
Anthracene	437	48			1-10,100	2
Fluoranthene	437	44			1-9,930	2
	12	100	0.34-11.4	2.06†		10
Pyrene	437	53			1-1,700	2
Chrysene	436	31			1-1,500	2
2,3-o-phenylenepyrene	437	2			17-102	2
	12	100	0.06-6.86	0.88†		10
Benzo(a)anthracene	437	27			1-1,500	2
Benzo(a)pyrene	437	5			1-490	2
	12	100	0.12-9.14	0.88†		10
3,4-benzofluoranthene	437	11			1-2,400	2
	12	100	0.06-9.14	1.47†		10
11,12-benzofluoranthene	438	8			1-379	2
	12	100	0.06-4.57	0.49†		10
Acenaphthylene	437	1			24-320	2
1,12-benzoperylene	437	2			12-133	2
	12	100	0.06-9.14	0.65†		10
1,2,5,6-dibenzanthracene	437	0.4			12-50	2
Isophorone	431	0	All samples < detection limit			2

Halogenated biphenyls

PCBs (Arochlor 1248)	431	0	All samples < detection limit			2
(Arochlor 1254)	107	39	0.0667-1,960	5.35		1
	431	0	All samples < detection limit			2
	31	100	0.15-3.6	0.84		3
	24	100	0.2-75	1.2		4
	14	93	0.35-23.1	4.0		6
(Arochlor 1260)	111	58	0.0468-433	4.18		1
	431	0	All samples < detection limit			2
	40	100	0.2-0.46	0.15		5
(decachlorobiphenyls)	74	100	0.11-2.9	0.99		7
(ref.std.unspecified)	9	100	0.36-7.60	1.20		8
(Arochlor 1016, 1221, 1232, 1242)	431 ea.	0	All samples < detection limit			2
PBB (polybrominated biphenyl)	210	0	All samples < detection limit			1

Chlorinated dibenzo-p-dioxins (CDDs) and chlorinated dibenzofurans (CDFs)§

Tetra-CDDs	2	(1)*		7.2 µg/kg		9
	2	(2)	0.138 and 0.222 µg/kg			11
Penta-CDDs	1	(0)	Sample < detection limit			9
Hexa-CDDs	2	(2)	0.3 and 2.1 µg/kg			9
	2	(2)	1.41 and 1.43 µg/kg			11
Hepta-CDDs	2	(2)	0.9 and 6.3 µg/kg			9
	2	(2)	9.4 and 7.6 µg/kg			11
Octa-CDDs	2	(2)	7.6 and 60 µg/kg			9
	2	(2)	50 and 60 µg/kg			11
Tetra-CDFs	2	(0)	Samples < detection limit			9
Penta-CDFs	1	(0)	Samples < detection limit			9
Hexa-CDFs	2	(0)	Samples < detection limit			9
Hepta-CDFs	2	(0)	Samples < detection limit			9
Octa-CDFs	2	(1)		1.3 µg/kg		9
2,3,7,8-TCDD	431	0	All samples < detection limit			2

Table 40. (continued)

Chemical	No. of sludges tested	% Occur- rences	Concentrations for samples testing positive, i.e., > detection limits			Ref.
			mg/kg (dry wt.) Range	Median	ug/L Range	
Halogenated aliphatics (short chain)						
Acrolein	431	0	All samples < detection limit			2
Acrylonitrile	155	61	0.0363-82.3	1.04		1
	436	1			5-290	2
bis(2-chloroethoxy)methane	431	0	All samples < detection limit			2
bis(chloromethyl)ether	431	0	All samples < detection limit			2
bix(2-chloromethyl)ether	431	0	All samples < detection limit			2
bis(2-chloroisopropyl)ether	431	0	All samples < detection limit			2
Carbon tetrachloride	436	4			5-3,030	2
Chlorodibromomethane	435	2			10-75	2
Chloroethane	436	7			5-71,000	2
2-chloroethyl vinyl ether	431	0	All samples < detection limit			2
Chloroform	436	24			1-366	2
Dichlorobromomethane	436	6			3-260	2
Dichlorodifluoromethane	436	9			2-4,300	2
1,1-dichloroethane	436	34			1-2,880	2
1,2-dichloroethane	436	10			1 10,000	2
1,1-dichloroethylene	436	3			1 14,000	2
1,2-dichloropropane	157	70	0.00243-66.0	0.464		1
	435	6			1-103	2
1,3-dichloropropane	158	26	0.209-309	3.08		1
1,3-dichloropropene	157	80	0.00203-1,230	3.42		1
	436	0.4			5-19	2
Hexachloro-1,3-butadiene	217	47	<10^{-4}-3.74	0.0355		1
	437	0.2			2,700	2
Hexachloroethane	216	61	0.00036-61.5	0.0199		1
	431	0	All samples < detection limit			2
Methyl bromide	431	4			33-30,000	2
Methyl chloride	436	6			12-6,100	2
Methylene chloride	436	73			1-10,500	2
Pentachloroethane	199	28	0.00025-9.22	0.030		1
1,1,2,2-tetrachloroethane	434	15			1-3,040	2
Tetrachloroethylene	128	73	<10^{-5}-0.122	0.00052		1
	436	40			1-2,800	2
Tribromomethane	436	0.2			5	2
1,1,1-trichloroethane	436	19			1-10,900	2
1,1,2-trichloroethane	434	4			1-2,100	2
1,2-trans-dichloroethylene	436	60			1-96,000	2
1,2,3-trichloropropane	141	48	0.00459-19.5	0.352		1
1,2,3-trichloropropene	137	48	<10^{-4}-167	1.14		1
Trichloroethylene	432	54			1-32,700	2
Trichlorofluoromethane	436	5			2-113	2
Vinyl chloride	435	8			8-62,000	2
Triaryl phosphate esters						
Cresyldiphenyl phosphate	238	3	0.607-179	18.9		1
Tricresyl phosphate	235	69	0.069-1,650	6.85		1
Trixylyl phosphate	236	68	0.0273-2,420	7.11		1

Table 40. (continued)

Chemical	No. of sludges tested	% Occur-rence*	Concentrations for samples testing positive, i.e., > detection limits			Ref.
			mg/kg (dry wt.)		µg/L	
			Range	Median	Range	

			Aromatic and alkyl amines§			
Benzidine	238	0.4	12.7			1
	431	0	All samples < detection limit			2
3,4-dichloroaniline	238	0	All samples < detection limit			1
3,3'-dichlorobenzidine	238	0	All samples < detection limit			1
	431	0	All samples < detection limit			2
p-nitroaniline	238	0	All samples < detection limit			1
N-nitrosodimethylamine	431	0	All samples < detection limit			2
	15	93	0.6-53 µg/kg	5.3 µg/kg		3
	11	82	0.5-93 µg/kg	2.5 µg/kg		4
N-nitrosodiethylamine	15	27	0.6-3.8 µg/kg	0.9 µg/kg		3
	11	54	0.9-12 µg/kg	1.7 µg/kg		4
N-nitrosodibutylamine	15	0	All samples < detection limit			3
	11	0	All samples < detection limit			4
N-nitrosopiperidine	15	0	All samples < detection limit			3
	11	9		2.9 µg/kg		4
N-nitrosopyrrolidine	15	13	2.0 and 4.2 µg/kg	2.7 µg/kg		3
	11	18	1.7 and 2.8 µg/kg	2.2 µg/kg		4
N-nitrosomorpholine	15	20	1.0-9.2 µg/kg	2.9 µg/kg		3
	11	54	1.3-2.9 µg/kg	1.8 µg/kg		4
N-nitrosodiphenylamine	431	0	All samples < detection limit			2
N-nitrosodi-n-propylamine	431	0	All samples < detection limit			2

			Phenols			
p-chloro-m-cresol	438	1			12-35	2
o-chlorophenol	231	9	0.0766-90.0	3.6		1
	438	2			11-72	2
m-chlorophenol	231	\7	0.123-93.3	0.891		1
p-chlorophenol	231	9	0.0277-90.0	3.28		1
o-cresol	231	6	0.177-183	2.05		1
2,4-dichlorophenol	230	7	0.209-203	4.76		1
	438	2			14-298	2
2,4-dimethylphenol	231	18	0.0899-86.7	2.19		1
	431	0	All samples < detection limit			2
4,6-dinitro-o-cresol	228	9	0.202-187	2.34		1
	431	0	All samples < detection limit			2
2,4-dinitrophenol	228	30	0.153-500	4.62		1
	431	0	All samples < detection limit			2
Hydroquinone	229	27	0.138-223	2.55		1
2-nitrophenol	431	0	All samples < detection limit			2
4-nitrophenol	431	0	All samples < detection limit			2
Pentachlorophenol	223	70	0.172-8,490	5.00		1
	438	14			10-10,500	2
Phenol	229	78	0.0166-288	2.00		1
	438	50			5-17,000	2
2,4,6-trichlorophenol	223	30	0.195-1,330	4.81		1
	438	0.4			11-16	2

			Chlorinated pesticides and hydrocarbons			
Aldrin	223	0	All samples < detection limit			2
	431	0	All samples < detection limit			2
	74	100	0.05-0.64	0.08		7
Chlordane	431	0	All samples < detection limit			2
	74	100	0.46-12	2.75		7
Dieldrin	221	28	0.000377-64.7	1.06		1
	431	0	All samples < detection limit			2
	40	?	<0.01-1.26	0.26		5
	14	93	0.04-2.2	0.16		6
	74	100	0.05-0.81	0.11		7

Table 40. (continued)

Chemical	No. of sludges tested	% Occur-rence*	Concentrations for samples testing positive, i.e., > detection limits			Ref.
			mg/kg (dry wt.)		μg/l.	
			Range	Median	Range	

Chlorinated pesticides and hydrocarbons (continued)

Endrin	223	0	All samples < detection limit			1
	431	0	All samples < detection limit			2
	74	100	0.11-0.17	0.14		7
Endrin aldehyde	431	0	All samples < detection limit			2
p,p'-DDD	221	48	0.00114-84.1	0.363		1
	431	0	All samples < detection limit			2
p,p'-DDE	219	92	0.00118-564	1.14		1
	443	0.2			10,000	2
	40	100	0.01-0.49	0.02		5
	74	0	All samples < detection limit			7
p,p'-DDT	219	95	<10⁻⁴-135	0.211		1
	431	0	All samples < detection limit			2
	74	100	0.06-0.14	0.09		7
Heptachlor	431	0	All samples < detection limit			2
Heptachlor epoxide	431	0	All samples < detection limit			2
	74	100	0.05-0.55	0.13		7
Lindane(γ-BHC)	221	17	0.00059-12.5	0.0746		1
	431	0	All samples < detection limit			2
	40	?	<0.01-0.93	0.18		5
	74	100	0.05-0.22	0.11		7
Methoxychlor	223	0	All samples < detection limit			1
2,4-D	223	25	0.000554-7.34	0.122		1
Toxaphene	431	0	All samples < detection limit			2
α-endosulfan	431	0	All samples < detection limit			2
β-endosulfan	431	0	All samples < detection limit			2
Endosulfan sulfate	431	0	All samples < detection limit			2
α-BHC	431	0	All samples < detection limit			2
β-BHC	431	0	All samples < detection limit			2
δ-BHC	431	0	All samples < detection limit			2

Miscellaneous compounds

1,2-diphenylhydrazine	431	0	All samples < detection limit			2
4-chlorophenyl phenyl ether	431	0	All samples < detection limit			2
4-bromophenyl phenyl ether	431	0	All samples < detection limit			2
Mercaptobenzothiazole	238	0	All samples < detection limit			1
Biphenyl	236	33	0.0437-1,730	8.61		1

Footnotes:

†A concentration range was given in Reference 10 for each of the 12 sludges tested. Therefore, the concentration reported as the median value for each PAH organic was obtained by taking the average of the high and low values to get an average concentration for each of the 12 sludges. These 12 averages were then used to report the range and median value in Table 40.

*"% occurrence" times the total "no. of sludges tested" equals the number of samples testing positive, i.e., having a concentration greater than the detection limit. Under "% occurrence", values given in parentheses are the number of samples which had a detectable concentration rather than a "percent" value.

§Note: Concentrations (on dry weight basis) for dioxins and furans and several alkyl amines are in μg/kg rather than mg/kg as for all other organics.

††References:

1. Jacobs and Zabik, 1983. (Various sludges from 204 Michigan WWTPs)
2. Burns and Roe, 1982. (Primary, secondary, and combined sludges from 40 POTWs)
3. Mumma et al., 1984. (Various sludges from 31 American cities)
4. Mumma et al., 1983. (Various sludges from 24 New York communities)
5. McIntyre and Lester, 1982. (Various sludges from 40 WWTPs in England)
6. Furr et al., 1976. (Various sludges from 14 American cities)
7. Clevenger et al., 1983. (Various sludges from 74 Missouri WWTPs)
8. Diercxsens and Tarradellas, 1983. (Various sludges from 9 Switzerland WWTPs)
9. Weerasinghe et al., 1985. (Two sludges from Syracuse, NY and Sodus, NY)
10. McIntyre et al., 1981. (Various sludges from 12 United Kingdom WWTPs)
11. Lamparski et al., 1984. (Two samples of Milorganite, one each produced in 1981 and 1982)

Table 41. Summary comparing the number of organic chemicals tested to the number of organics not detected in sewage sludges or found in 10, 50 or 90% of the sludges.

Reference[*]	No. of organic chemicals tested	No. of organics undetected in all samples tested	No. of organic chemicals with occurrence:		
			>10%	>50%	>90%
Phthalate esters					
1 (204 WWTPs)	6	0	6	3	0
2 (40 POTWs)	6	0	4	1	1
Monocyclic aromatics					
1 (204 WWTPs)	23	12	8	3	0
2 (40 POTWs)	12	3	7	3	1
Polynuclear aromatics (PAH)					
1 (204 WWTPs)	1	0	1	1	0
2 (40 POTWs)	18	1	9	2	0
10 (12 WWTPs)	6	0	6	6	6
Halogenated biphenyls					
1 (204 WWTPs)	3	1	2	1	0
2 (40 POTWs)	3	3	0	0	0
3 (31 WWTPs)	1	0	1	1	1
4 (24 WWTPs)	1	0	1	1	1
5 (40 WWTPs)	1	0	1	1	1
6 (14 WWTPs)	1	0	1	1	1
7 (74 WWTPs)	1	0	1	1	1
8 (9 WWTPs)	1	0	1	1	1
Dioxins and furans					
2 (40 POTWs)	1	1	0	0	0
9 (2 WWTPs)	(not enough samples tested to suggest % occurrence)				
11 (2 samples of Milorganite)	(not enough samples tested to suggest % occurrence)				
Halogenated aliphatics					
1 (204 WWTPs)	10	0	10	5	0
2 (40 POTWs)	32	7	9	3	0
Triaryl phosphate esters					
1 (204 WWTPs)	3	0	2	2	0
Aromatic and alkyl amines					
1 (204 WWTPs)	4	3	0	0	0
2 (40 POTWs)	5	5	0	0	0
3 (15 WWTPs)	6	2	4	1	1
4 (11 WWTPs)	6	1	4	3	0
Phenols					
1 (204 WWTPs)	12	0	6	2	0
2 (40 POTWs)	11	5	2	0	0
Chlorinated pesticides and hydrocarbons					
1 (204 WWTPs)	9	3	6	2	2
2 (40 POTWs)	19	18	0	0	0
5 (40 WWTPs)	3	0	1	1	1
6 (14 WWTPs)	1	0	1	1	1
7 (74 WWTPs)	8	1	7	7	7
Miscellaneous compounds					
1 (204 WWTPs)	2	1	1	0	0
2 (40 POTWs)	3	3	0	0	0
TOTALS:	219	70	102	53	26

[*]Reference number used refers to the same references as used in Table 40.

[†]Waste Water Treatment Plant (WWTP); Publicly Owned Treatment Plant (POTW).

plant system by application to land. Table 42 summarizes that part of the analysis data from Table 40 which had median concentration values. This summary suggests that about 90% of the organics in sludges will be present at concentrations less than 10 mg/kg. About 10% of the organics tested had median concentrations of 11–100 mg/kg, and only one organic had a median value of >100 mg/kg (Table 42).

To put potential organic chemical loadings into perspective, one can make a comparison with agricultural pesticides. Many pesticides used today are organic chemicals which are added to soil-plant systems at rates of 0.2–4.0 kg of active ingredient per hectare. Assuming an agronomic rate of sludge application of 10 mt/ha (dry weight basis), the organic chemical loadings expected for organic concentrations in sludges of 1, 10, and 100 mg/kg are 0.01, 0.1 and 1.0 kg/ha. At rates used to reclaim drastically disturbed land, 100 mt/ha, the organic loading for sludges containing 1, 10, and 100 mg/kg organic concentration would be 0.1, 1.0, and 10 kg/ha, respectively. For agronomic rates organic chemical concentrations of sludges approaching 100 mg/kg must be viewed as potentially having an impact on the soil-plant system, depending on the chemical/ toxicological properties of that organic. At high sludge rates (e.g., 100 mt/ha), concentrations approaching 10 mg/kg in sludge could be expected to add amounts of an organic comparable to quantities of pesticides added in agricultural operations.

Table 42. Summary showing the distribution of median dry matter concentrations for data reported in Table 40.[a]

Chemical group	No. of organic chemicals tested	No. of organic chemicals tested having median concentrations in sludges (mg/kg, dry wt. basis):				
		ND[†]	<1	1-10	11-100	>100
Phthalate esters	6	0	0	1	4	1
Monocyclic aromatics	23	12	5	2	4	0
Polynuclear aromatics (PAH)	7	0	4	2	1	0
Halogenated biphenyls	9	1	3	5	0	0
Dioxins and furans	(inadequate data; all concentrations reported in Table 40 significantly <1 mg/kg or 1000 µg/kg)					
Halogenated aliphatics	10	0	6	4	0	0
Triaryl phosphate esters	3	0	0	2	1	0
Aromatic & alkyl amines	16	6	9	0	1	0
Phenols	12	0	1	11	0	0
Chlorinated pesticides and hydrocarbons	21	4	14	3	0	0
Miscellaneous	2	1	0	1	0	0
TOTALS:	109	24	42	31	11	1

[a]Summary does not include data from Burns and Roe (1982) which was reported on a wet basis without median values provided. Also note that median values used are only for those samples having detectable concentrations and are not true median values, which would be lower if all "ND" samples were included as zeroes.

[†]ND = organic was "not detected" in any sludge samples tested

Based on the prevalence of organics in sludges and potential loadings to soils, agronomic or environmental risk due to the application of domestic sewage sludge to agricultural soils appears to be minimal. In addition, many organics will be bound by soil organic matter and biologically degraded by soil microorganisms (Kaufman, 1983). However, persistent compounds like PCBs and the chlorinated pesticides could accumulate in soils from repeated sludge applications and can be a concern for food crop production.

TRACE ORGANICS IN SOILS

Limited information is available regarding residual effects of sludge organics in soils. Monitoring for 22 persistent organics in unamended and sludge-amended soils (Baxter et al., 1983) showed trace levels of chlordane (<0.12 mg/kg), dieldrin, p,p'-DDE and PCBs present in untreated and treated soils. None of the other 22 organics were detected in any of the soil samples. The authors concluded that sludge applications had not measurably increased the level of persistent organics above the levels normally found.

In a Michigan study (Singh, 1983), sludge-treated and untreated soils were collected from 15 sites around the state and analyzed for 10 organic compounds: benzene, trichloroethylene, tetrachloromethane, PCBs, pentachloronitrobenzene, pentachlorophenol, chlorpyrifos, di-n-butylphthalate, bis(2-ethylhexyl) phthalate, and toxaphene. All results for soil analyses were reported as "none detected" except at two sites: PCBs were found at 0.8 mg/kg for untreated and sludge-treated soils at one site, and 0.02 mg/kg pentachlorophenol was detected in sludge-treated but not in control soils at another site.

While these two studies suggest that sludge organic chemical loadings to soils will result in little or no residues in soils receiving sludges, additions of persistent organics can potentially be a concern. Two food processing companies were contacted to determine what level(s) of organic residues in soils they use to reject fields for use in growing vegetable crops. One company indicated that concentrations above 0.1 mg/kg of aldrin/dieldrin, chlordane, toxaphene, or lindane in mineral soils would be of concern. For muck soils, 1.6 mg/kg of aldrin/dieldrin or 0.5 mg/kg of chlordane, toxaphene or lindane could be tolerated. A second vegetable crop processing company provided the guideline information in Table 43.

To assess the potential impact of sludge organic loadings to agricultural soils, the theoretical residue levels can be determined. Using the highest median concentration for aldrin or dieldrin from Table 40 of about 1 mg/kg and assuming an agronomic rate of sludge is applied (10 mt/ha) for 10 years, the total amount of aldrin or dieldrin added to a soil would be: 0.01 kg/ha x 10 (yr) = 0.1 kg/ha. To determine what the soil residue concentration would be, one can assume an average bulk density for soil of 1.3 g/cm^3 (1,300 kg/m^3) and a 20 cm depth of mixing, so one hectare (10,000 m^2) of soil would weigh 2,600,000 kg (10,000

Table 43. Guidelines used by one food processing company for interpreting
the significance of residues in soils being considered for
growing root crops.

Significance	Range of soil residues (mg/kg)		
	Aldrin/Dieldrin	DDT	Diuron*
Suitable for planting†	0-0.1	0.75	0.3
May be planted, but crop must be analyzed before acceptance	0.1-0.2	0.75-1.5	0.3-0.5
Do not plant	>0.2	>0.5	>0.5

*Regardless of residues present, beets and carrots must not be planted in
soil which has received (a) an improper application of diuron, or (b) an
application of diuron for which the minimum treatment-to-planting interval has not expired.
†Plant carrots in least-contaminated soil.

m^2 x 0.2 m x 1,300 kg/m^3). Assuming no loss of the organic
chemical applied, the soil residue concentration would be:
0.1 kg (or 100,000 mg) of organic/ha ÷ 2,600,000 kg of soil/ha
= ~0.04 mg/kg.

   Under these conditions, some margin of safety would still be
provided relative to the 0.1 mg/kg guideline level (Table 43)
used for most sensitive root crops. However, if the sludge
organic concentration was 10 mg/kg instead of 1 mg/kg, then the
same sludge loading would give a soil residue level of 0.4 mg/kg
and cause such a soil to be excluded for growing vegetable root
crops. But a sludge with 10 mg/kg of aldrin or dieldrin would
still be acceptable if 1 ton per hectare per year was applied for
10 years instead of 10 ton per hectare per year. Therefore, the
total amount applied to a soil is the critical factor rather than
the concentration in the sludge. Again, as noted above, these
examples assume no loss of the organic chemical by volatilization, degradation, etc. from soil.

EXTRACTION/LEACHING PROCEDURES

   Under the Amendments to the Resource Conservation and Recovery
Act of 1985 [Hazardous Waste Management System; Definition of
Solid Waste; Final Rule (40 CFR Parts 260, 261, 264, 265 and
266), January 4, 1985], the U.S. EPA was directed to improve the
ability to characterize hazardous waste. The Extraction
Procedure Toxicity Characteristic (EPTC), or EP toxicity test,
currently used entails a leaching test to measure the tendency of
a waste to leach, coupled with extract concentrations above which
the waste is to be regulated, and defined as a hazardous waste.
This test was developed on the premise that a potentially
hazardous industrial waste might be sent to a sanitary landfill,
resulting in a high potential for groundwater contamination. The
constituents currently included as part of this test were those
for which National Interim Primary Drinking Water Standards have
been established. These standards addressed 8 inorganics and 6
organic compounds (2,4-dichlorophenoxyacetic acid, endrin, lindane, methoxychlor, toxaphene, and 2,4,5-trichlorophenoxyacetic
acid).

As part of the effort to improve the characterization of hazardous waste, EPA will be proposing a revised test (Friedman, 1985) that would expand the list of organic compounds tested to 44 and modify the procedure itself to the Toxicity Characteristic Leaching Procedure (TCLP). If the extractant concentration from the TCLP is above the maximum threshold limit for any of the 8 inorganic or 44 organic chemicals, the waste is defined as hazardous. Each municipality that produces sewage sludge must make the determination whether or not their sludge is hazardous. This determination can be based upon their knowledge of their sludge or they may choose to use the TCLP to help them make it. The EPA believes that the way to determine if a municipal sludge is hazardous is to determine whether or not its extract concentrations exceed the maximum threshold limits. However, many have argued that a testing procedure based on the worst-case scenario in which large quantities of sludge are disposed of in a landfill has little relevance to assessing any potential hazard from recycling low rates of sludge to land.

The U.S. EPA is currently having eight sewage sludges tested with the new TCLP procedure. The sludges were selected to include purely domestic sewage sludge as well as sludges expected to have high concentrations of contaminants from industrial sources. While none of the sludges appear to have TCLP extract concentrations that exceed the threshold limits, results are too preliminary to know for sure. Therefore, the impact of changing from the EPTC to the TCLP on sludge application programs is too early to ascertain.

MUTAGENICITY TESTING OF SLUDGES

A number of studies have recently reported results of mutagenicity tests on extracts of sludge (Babish et al., 1983; Boyd et al., 1982; Hang et al., 1983; Hopke and Plewa, 1984; Hopke et al., 1982). While most sludge extracts tested by Babish et al.(1983) were mutagenic by the Ames test (Ames et al., 1975) many foods, drinking water, and other substances in our environment also test positive for mutagenic activity (Loper, 1980; Mast et al., 1984; Nagao et al., 1979; and Salmeen et al., 1985). Ames (1983) has also indicated that "the human diet contains a great variety of natural mutagens and carcinogens, as well as many natural antimutagens and anticarcinogens". Thus, one must use extreme care in interpreting mutagenic tests of sludge extracts to keep them in perspective with the presence of mutagenic constituents in all parts of our environment.

In addition to the Ames Salmonella assay, plant test systems have been used to investigate the mutagenic activity of sewage sludges (Hopke and Plewa, 1984; Hopke et al., 1982). Mutagens present in sludge-amended soil can be transported into a crop plant and induce genetic damage in germ cells; however, no mutagenicity occurred in the kernels from corn grown on sludge-amended soil nor were mutagens transferred from the sludge to soil or surface waters.

These studies imply that the chemicals causing mutagenicity are trace organics, but the specific chemicals responsible for the mutagenic effects have not been identified. Another difficulty in interpreting these bioassay results is that agricultural soils can exhibit mutagenic response without sewage sludge amendments (Boyd et al., 1982; Brown et al., 1985; and Hopke et al., 1982). Therefore, results of these mutagenicity tests are not easy to put into perspective (Davis et al., 1984; Dean and Suess, 1985). However, the data cited suggest that mutagen activity is greater for sludges generated by municipalities which have more industrial dischargers.

While mutagens present in sludges were shown to degrade relatively rapidly (e.g., within 2-3 weeks) in one sludge-amended soil (Angle and Baudler, 1984), recent work at Pennsylvania State University indicates that the loss of mutagenic activity may take as long as one growing season for other sewage sludge/soil mixtures (Baker et al., 1985). How well results from these laboratory incubations are duplicated under field conditions is untested and still unknown.

Due to the large number of organic chemicals which can be present in sewage sludge, a short-term bioassay offers the advantage of testing for potential biological toxicity inherent in a sludge (or other waste) containing a complex mixture of chemicals (Brown et al., 1982). For example, Peters (1985) used the Ames test to screen 38 Pennsylvania sewage sludges potentially containing harmful trace organics, and Brown et al. (1982) used the Ames test plus two other bioassays to examine the acute toxicity of ten hazardous wastes.

Using a bioassay test for identifying a sludge contaminated with an organic chemical(s) could provide an additional degree of safety in managing sewage sludge applications to agricultural and forest soils. To be useful, however, bioassay test results for sludges must be correlated to mutagenic activity or biological toxicity of soil/sludge mixtures in the field. Based on the generally low concentrations of trace organics in sludges and the low rates of sludge (e.g., agronomic) typically utilized, the probability of any transfer of mutagenic activity to animals or humans as a result of sludge application to land is very low.

FATE OF TRACE ORGANICS ADDED TO SOIL-PLANT SYSTEMS

Potential health hazards associated with organic chemical residues in sludge applied to land have been discussed in several review articles (Chaney, 1984; Dacre, 1980; Davis et al., 1984; Kowal, 1983, 1985; Majeti and Clark, 1981; and Pahren et al., 1979).

Principal pathways by which organics could be transferred to humans from sludge-amended soils were listed by Dean and Suess (1985):

1. Uptake by plant roots in sludge-treated soil, transfer to edible portions of plants, consumption by humans;

2. Direct application to edible parts of plants as sludge, or as dust or mud after sludge is mixed with the soil, consumption by humans;

3 Uptake via plants used as feed or fodder for animals, transfer to animal food products, consumption by humans;

4. Direct ingestion of soil and sludge by grazing animals and transfer to animal food products, consumption by humans;

5. Direct ingestion of sludge contaminated soil by children; an abnormal behavior called "pica".

Two other possibilities might be included with this list:

6. Surface runoff/erosion to streams or rivers used as a source of drinking water downstream, and

7. Leaching to a groundwater aquifer used as a source of drinking water.

These pathways have all been demonstrated but are not equally important. Indeed, Pathway 4, which does not go through plants, is the only one by which organic pollutants have been traced from sludge to animal products (Chaney, 1984).

While plant contamination can occur (as discussed later), soil residue levels necessary for this to happen are usually higher than would be anticipated from low application rates of non-industrialized sewage sludges. In addition, soil incorporation of organics, like PCBs for example, can greatly reduce plant "uptake" of the chemical (Harms and Sauerbeck, 1983). Lindsay (1983) also reported that several recalcitrant organics are so strongly bound to soil and sludge as to be almost totally unavailable for plant uptake.

Trace organics may biomagnify. For example, detritus eating insects were found to contain 1.3 x the soil concentration of PCB, which could lead to further bioconcentration in insect-eating birds (Davis et al., 1984). As with metals, trace organics may accumulate in animal food products following direct sludge ingestion during grazing. The problem is particularly important for dairy cows since milk is the animal product most likely to be influenced by organic contaminants in sludge applied to land (Dean and Suess, 1985), although management practices can significantly reduce this possibility.

The potential exists for direct ingestion of organics, especially by children through the phenomenon of pica (Pathway 5) if sludge was used to fertilize home gardens. Dean and Suess (1985) concluded, however, that this is likely to be a minor or insignificant route of exposure, as is inhalation of dust or vapors. As with Pathway 1, significant human consumption of sludge organics by human management (e.g., culinary procedures like cleaning and peeling of root crops that tend to accumulate lipophilic substances) would seem most unlikely (Naylor and Loehr, 1982b).

Bioaccumulation factors (i.e., ratio of an organic in plant or animal tissue to concentrations in soil) are available for very few compounds. For plants, the factor (when known) is almost always <1 and usually <0.1 (Overcash et al., 1985) and for animal products (e.g., milk) estimates of 0.7 for PCB (Fries, 1982) and 0.5 for dieldrin (Lindsay, 1983) have been made. Field data are largely non-existent, but Baxter et al. (1983) reported no plant uptake of 22 persistent organics from land amended with Denver, Colorado, Metro sludge.   Also, no increases in persistent trace organics content of the fat tissue content of cattle grazing a sludge application site were observed.

If ingested, organics present in sludge or soil can be bio-available (Chaney, 1984; McConnell et al., 1984). Jelinek and Braude (1977) found an increased content of PCBs in the milk fat of cattle fed green forages and roughages grown on sludge-treated land. This prompted the U.S. FDA to recommend a maximum permissible content of not more than 10 mg/kg PCBs in sludges used on agricultural lands (Braude et al., 1975). Therefore, absorption of ingested sludge organics can only be prevented by limiting their concentration in sludges or avoiding direct ingestion until compounds have been degraded or dissipated.

Assimilative Pathways Within the Soil-Plant System

Organic compounds may undergo a variety of chemical and biological processes when applied to a soil or soil-vegetation system. The various assimilative pathways have been discussed by several authors (Davis et al., 1984; Kaufman, 1983; Lue-Hing et al., 1985; Overcash, 1983) and include:

1. adsorption onto soil and its constituents;

2. volatilization;

3. degradation (microbial, chemical, photolysis);

4. leaching to groundwaters and runoff/erosion to surface waters;

5. plant retention (contamination vs. uptake and translocation); and

6. macro- and micro-fauna uptake (bioaccumulation by insects, grazing animals).

While research on the fate of sludge organics in soils is limited, the behavior of organics in soil, has been extensively studied, particularly for agricultural pesticides, (e.g., Goring and Hamaker, 1972; Guenzi et al., 1974) and with non-agricultural chemicals in the petroleum industry (API, 1983). In general, trace organics are strongly adsorbed to soils and its constituents, especially soil organic matter. Thus, leaching and plant uptake are usually very limited. Some runoff/erosion may occur for organics firmly bound to soil particles or debris, but this can be minimized by using good soil and water conservation practices at sludge application sites.

Some trace organics (notably, PCBs, lindane, dieldrin) are known to volatilize readily when surface applied, although soils and sludge itself can drastically reduce these volatizilation losses. Some organics are recalcitrant to microbial degradation, but most are expected to degrade. Of concern are the persistent organics and some of the readily degraded components that can break down to toxic metabolites. Pathways of most interest are plant uptake/contamination, degradation, volatilization, and leaching.

Plant Uptake/Contamination

Chaney (1984) provided a good discussion about the uptake of organics by plants. Because this summary seemed more appropriate than others (Davis et al., 1984; Harms and Sauerbeck, 1983; Kaufman, 1983; Lue-Hing et al., 1985; and Overcash, 1983) and is not readily available, much of his discussion was used for this section, sometimes verbatim. Readers are encouraged to review these other references, however, for a more complete understanding of plant uptake/contamination.

Trace organics can enter edible parts of plants by two processes: 1) uptake from the soil solution, with translocation from roots to shoots, or 2) absorption by roots and shoots of volatile organics from the soil. Systemic pesticides are applied to the soil, then absorbed and translocated to the plant leaves. These kinds of compounds are quite water soluble and would probably not appear in wastewater treatment sludges at appreciable levels.

Lipophilic, halogenated organics are examples of water insoluble compounds which may contaminate plants by vapor absorption from the soil air or the organic-enriched air near the soil surface. Beall and Nash (1971) developed a method to discriminate between movement of an organic through the plant vascular system (uptake-translocation) vs. vapor phase movement. They found soybean shoots were contaminated by soil-applied dieldrin, endrin, and heptachlor largely by uptake-translocation, while vapor transport predominated for DDT and was equal to uptake-translocation of endrin. Using this method, Fries and Marrow (1981) found PCBs reached shoots via vapor transport, while the less volatile PBBs did not contaminate plant shoots by either process (Chou et al., 1978; Jacobs et al., 1976).

Root crops are especially susceptible to contamination by the vapor-transport route. Carrots have a lipid-rich epidermal layer (the "peel") which serves as a sink for volatile lipophilic organics. Depending on the water solubility and vapor pressure of the individual compound, it may reside nearly exclusively in the peel layer of carrots, or penetrate several millimeters into the storage root (Fox et al., 1984; Iwata and Gunther, 1976; Iwata et al., 1974; Jacobs et al., 1976; Landrigan et al., 1978; Lichtenstein et al., 1964, 1965; Lichtenstein and Schulz, 1965).

Carrot cultivars, however, were found to differ in uptake and in the distribution of the chlorinated hydrocarbon pesticides endrin and heptachlor between peel vs. pulp (Hermanson et al., 1970; Lichtenstein et al.). Other root crops (sugar beet, onion, turnip, rutabaga) are much less effective in accumulating lipophilic organics in their edible roots, possibly because the surface of the peel is lower in lipids (Chou et al., 1978; Fox et al., 1964; Lichtenstein and Schulz, 1965; Moza et al., 1976, 1979).

Based on carrot accumulation of volatile chlorinated hydrocarbon pesticides, Iwata et al. (1974) evaluated PCB uptake by carrots in the field from a low organic matter (0.6%) sandy soil, which represents a worst-case assessment, where 100 mg/kg Arochlor 1254 was applied in the surface 0-15 cm soil. For the environmentally persistent 5 and 6 chlorine isomers, unpeeled fresh carrots contained PCB at about 4.9% of the soil level. Peeling removed 14% of the carrot fresh weight and 97% of the PCB residue, so peeled fresh carrots contained PCB at only 0.16% of the soil PCB level.

The level of chlorinated hydrocarbon in carrots is also sharply reduced by increased organic matter in soil. The increased organic matter adsorbs lipophilic compounds and keeps them from being released to the soil solution or soil air (Chou et al., 1978; Filonow et al., 1976; Strek et al., 1981; Weber and Mrozek, 1979). For example, the additions of sewage sludge can increase the ability of soils to adsorb PCBs (Fairbanks and O'Connor, 1984) and may completely counteract the low levels of PCBs added by the sludge.

The residue of PCBs in waste materials such as municipal sludge can be depleted of the more volatile and more easily biodegraded lower chlorinated compounds. Because plant contamination via volatilization is much less for higher chlorinated compounds than for the more volatile lower chlorinated compounds at equal soil concentrations (Fries and Marrow, 1981; Iwata and Gunther, 1976; Moza et al., 1976, 1979; Suzuki et al., 1977), the lack of plant contamination from sludge-applied PCBs is not unexpected. For example, in a study by Lee et al. (1980), a sludge containing 0.93 mg/kg PCBs was applied at a rate of 112 dry mt/ha, yet "no PCBs were detected in the sludge grown carrots". Since other root crops are not nearly as good PCB accumulators as carrot (Moza et al., 1979), remarkably low potential human PCB exposure would be predicted for recommended sludge utilization practices.

Risk from polycyclic aromatic hydrocarbons (PAHs), some of which are carcinogenic (e.g., benzo(a)pyrene) has also been assessed. Carrot roots (but not mushrooms) accumulated many PAHs from compost-amended soils (Borneff et al., 1973; Ellwardt, 1977; Linne and Martens, 1978; Müller, 1976; Siegfried, 1975; Siegfried and Müller, 1978; Wagner and Siddiqi, 1971). The level of 3,4,-benzypyrene in carrot roots declined with successive cropping of compost amended soils. Harms and Sauerbeck (1983) also found PAH contamination of potato tubers and the roots of radishes and carrots where direct contact with the soil allowed transfer of these organics.

Concentrations in the above-ground parts of plants were, however, low.

Nitrosamines are another group of organics which have been found in sewage wastes (Green et al., 1981; Yoneyama, 1981). Although accumulated from nutrient solution and soil by plants (Brewer et al., 1980; Dean-Raymond and Alexander, 1976), nitrosamines appear to be rapidly degraded in soils and plants. Research on N-nitrosodimethylamine and N-nitrosodiethylamine found plant uptake could occur initially, but these compounds were rapidly degraded (Dressel, 1976a, 1976b). While traces of nitrosamines are found in nitroanaline based herbicides no detectable nitrosamine was found in soybean shoots due to their rapid degradation (Kearney et al., 1980b). An International Union for Physics and Chemistry (IUPAC) committee assessed the environmental consequences of these trace nitrosamines, and found no risk to the food chain (Kearney et al., 1980a).

Many other carcinogenic or toxic compounds could be present in sludges and contaminate the food chain through plant uptake or volatile contamination of crop roots or shoots. While information on these other organics is limited, two data bases are available which consider plant uptake of organic molecules. PHYTOTOX deals with the direct effect of exogenously supplied organic chemicals on the growth and development of terrestrial plants (Royce et al., 1984). As of July 1985, 9,800 papers had been included with data on 3,500 chemicals and 700 species (Rygiewicz, 1986, Personal communication). This data base is now available through a private service (Fein-Marquart Associates, 7215 York Rd., Baltimore, MD 21212). The second data base (UTAB) contains information pertaining to the Uptake, Transport, Accumulation and Biotransformation of organic compounds by vascular plants. This data base includes 3,900 papers, with information about 700 chemicals and 250 species and is available through the University of Oklahoma (John Fletcher, Dept. of Botany, Univ. of Oklahoma, Norman, OK 73019). These data bases offer the opportunity to evaluate basic research on the uptake of organics by plants which may help to understand the effects of sludge applied organics.

Degradation

Degradation of organic chemicals in soil may occur by chemical, photochemical, or biological processes. Degradability of a compound depends on its chemical structure, some being rapidly decomposed while others are relatively recalcitrant to degradation. Biodegradation can occur in microbial cells, in the soil solution by chemical mechanisms, or by extracellular enzymes sorbed to soil particles (Kaufman, 1983).

Often, soil microbes capable of degrading a compound proliferate in soil, and the effective population may remain several years after the last treatment. Maintaining a supply of biodegradable organic matter in soils receiving wastes would likely provide a higher population of diverse microbes capable of degrading more kinds of trace organics. Microbes may utilize a particular organic as an energy source, or may cometabolize it with other normal metabolic processes. Although the kinds of organisms and even types of enzymes involved in biodegradation are known for some pesticides and other organics, little is known about most of the organics found in wastes like sewage sludges (Kaufman, 1983).

Microbiological as well as chemical reactions are usually acting simultaneously. Chemical reactions (abiotic routes) are a part of the overall measure of organic compound decomposition. Two typical reactions are hydrolysis and neutralization of the parent organic species, but such reactions typically leave the bulk of the parent structure still intact (Overcash, 1983). Soil factors known to affect chemical degradation of organics include temperature, aeration, microbial populations, pH, organic matter, clay, cation exchange capacity, and moisture (Kaufman, 1983).

The action of sunlight may chemically alter and degrade organic chemicals in the environment. The importance of photo-chemical reactions to the degradation of waste organics applied to land will depend largely upon the mode of application and soil incorporation. Sludge organics can be subjected to some photoly-tic action during the time they are on exposed soil surfaces following a surface application (Kaufman, 1983). Under these conditions an organic compound may be degraded via photolytic mechanisms. Phenolics and polynuclear aromatics are two groups that readily undergo such reactions (Overcash, 1983). Photolytic degradation will be nonexistent, however, when sludges are incor-porated into the soil since sunlight does not penetrate the soil surface (Kaufman, 1983).

Following an extensive literature search concerning the decomposition of specific organics in the terrestrial environ-ment, Overcash (1983) concluded that very few organic compounds can be said to be non-degradable. Considering the long time periods typical in soil systems, only two classes of compounds were regarded as nondegradable based on available terrestrial research information: (1) synthetic polymers manufactured for stability, and (2) very insoluble large molecules, e.g., 5-10 chlorinated biphenyls (Overcash, 1983).

Other organics will have varying decomposition half-lives or persistence in soils. Overcash (1983) provided examples of half-life ranges for several organic chemicals (Table 44) and Kaufman (1983) listed the relative persistence for several organic chemi-cal classes (Table 45). Tabak et al. (1981) compared the rela-tive biodegradation of organic priority pollutants with a static culture flask procedure. While their decomposition results may not be directly extrapolated to degradation of organics in the soil, the relative degree of biodegradation may prove to be simi-lar in soils. Significant biodegradation was found for phenolic compounds, phthalate esters, napthalenes, and nitrogenous orga-nics; variable results were found for monocyclic aromatics, poly-cyclic aromatics, polychlorinated biphenyls, halogenated ethers, and halogenated aliphatics; and no significant biodegradation was found for organochlorine pesticides.

Table 44.   Illustrative range of decomposition half-life for organic
compounds.*

Compound	Approximate half-life
Aminoanthroquinone dyes	100-2,200 days
Anthracene	110-180 days
Benzo(a)pyrene	60-420 days
Di-n-butylphthalate ester	80-180 days
Nonionic surfactants	300-600 days
2,4-methyaniline	1.5 days
n-Nitrosodiethylamine	40 days
Phenol	1.3 days
Pyrocatechin	12 hours
Cellulose	35 days
Acetic acid	5-8 days
Hydroquinone	12 hours

*From Overcash (1983), p. 211

Table 45.   Relative persistence and initial degradative reactions of
nine major organic chemical classes.*

Chemical class	Persistence	Initial degradative process
Carbamates	2-8 weeks	Ester hydrolysis
Aliphatic acids	3-10 weeks	Dehalogenation
Nitriles	4 months	Reduction
Phenoxyalkanoates	1-5 months	Dealkylation, ring hydroxylation or oxidation
Toluidine	6 months	Dealkylation (aerobic) or reduction (anaerobic)
Amides	2-10 months	Dealkylation
Benzoic acids	3-12 months	Dehalogenation or decarboxylation
Ureas	4-10 months	Dealkylation
Triazines	3-18 months	Dealkylation or dehalogenation

*From Kaufman (1983), p. 119.

Volatilization

Vapor movement of organics (i.e., diffusion and volatilization) are important factors affecting the distribution and persistence of some organic chemicals in soil. An estimate of potential volatility can be obtained from the ratio of water solubility to vapor pressure, which indicates the proportion of an organic in the vapor phase. This ratio is only a guide, however, since adsorption of the organic in the soil will decrease the amount present in the vapor phase (Kaufman, 1983).

An organic spread on the surface of or injected into a soil with sludge will partition between the gas and liquid phases to exert a vapor pressure. The conditions of the soil and the application technique used, as well as the inherent organic compound volatility, are important factors in quantifying how an organic compound might be lost through volatilization. The level of vapor pressure at which volatile losses are known to be significant is usually taken at 5×10^{-6} mm Hg at 25°C. However, vapor pressure alone may be misleading because highly volatile organics like toluene are prevalent in municipal sludges, even after opportunities have occurred for volatile loss during wastewater treatment (Overcash, 1983).

Volatilization losses were considered as significant processes of organic chemical removal when wastewaters are applied to land (Chang and Page, 1984; Jenkins et al., 1983). Jenkins et al. (1983) stated "as a rule the higher the vapor pressure the lower the water solubility, the higher the Henry's law constant and the higher the removal rate by volatilization." Once the organic reaches the soil, the actual volatilization loss of trace organics from the soil will depend on factors affecting the movement of the organics to the soil surface and its dispersion into the air (Chang and Page, 1984).

For soil-applied pesticides, the vapor density was found to be the main factor controlling volatilization (Farmer et al., 1972). Other factors which affect volatilization include soil pesticide concentration, temperature, rate of air movement over the soil surface, and soil water content (Farmer et al., 1972; Igue et al., 1972). More recently Jury et al. (1983, 1984a,b) have used benchmark properties of vapor density and solubility in water in a mathematical model to determine the relative volatility of a specific soil-applied organic.

As with earlier work done with pesticides, how well research results will predict volatilization losses for the same organics applied to soils as part of a sludge matrix is unknown. Research reported by Fairbanks and O'Connor (1984) indicate that sludge additions to soil can decrease volatilization losses of PCBs, so the sludge matrix could be expected to have some effect. Nevertheless, models and research data which apply to soil-applied pesticides provide a good "point of departure" for understanding potential volatilization losses of pesticides and other organics added by sludge applications.

Leaching

The downward movement of an organic chemical is largely governed by sorption and biodegradation. At least two steps are involved in the leachability of an organic chemical in soil: (1) entrance of the compound into solution, and (2) adsorption of the compound to soil surfaces (Kaufman, 1983). Partitioning between the adsorbed and soil solution phases may occur immediately upon application to the soil or may be delayed until the organic separates from the waste medium. At the same time, decomposition reactions can influence the amount of a particular organic compound which may reside on the soil/waste phase or in the soil solution (Overcash, 1983).

Therefore, the inherent persistence of each chemical in soil will affect whether any mobile chemical might pollute groundwater. The half-life of many organic chemicals in soil is sufficiently short to make it highly unlikely that the chemical would ever reach the water table under ordinary field leaching conditions (Kaufman, 1983). Overcash (1983) also concluded that for most sludge application sites where normal application rates and management techniques are used, leaching of organics is probably negligible. Nevertheless, leaching of organic chemicals can occur, as evidenced by recent discoveries of pesticide contamination in groundwaters, and should not be overlooked.

Effects of Sludge Properties

Few studies have considered the effect of sludge on the assimilative pathways of adsorption, volatilization, and degradation. Since organics typically associate with the organic fraction of soils, one might expect even greater retention of trace organics in amended vs. unamended soils as was shown by Fairbanks and O'Connor (1984) for di-3-(ethylhexyl)phthalate (DEHP), PCBs, and two herbicides. The greater adsorption in sludge-amended soils should reduce contaminant mobility and plant availability, and data did show that volatilization of PCBs from sludge-amended soils was significantly reduced (Fairbanks and O'Connor, 1984).

Sludge additions may also affect organic contaminant degradation. The increased microbial activity found in sludge-amended soils suggests that previous sludge applications cause a preconditioning with respect to microbes and/or enzymes which may increase organic degradation by cometabolism (Fairbanks and O'Connor, 1984). The degree and duration of sludge effects on trace organic behavior are influenced by type and concentration of compound, incubation time, sludge rate, and soil type.

Most experiments designed to determine the effects of sludge or organic behavior have used "spiked" systems in which the target organic is added as reagent grade chemical to soil or soil-sludge systems. Preequilibrium of the target organic with sludge has been minimal. Thus, most data generated to date are tainted by limitations similar to the early "mineral salt" work with metals. Organics indigenous to sludge may have drastically different properties with respect to their fate in soils compared to these same trace organics added to the soil alone or in combination with sludge. Research is needed with selected sludges to study the assimilative pathways of specific organics (indigenous to these sludges) compared to results for comparable amounts of the same organic added to the same soil in the absence of the sludge matrix. Correlations between controls and sludge-treated soils could then be used to predict the fate of other sludge-applied organics when actual field data are unavailable.

Utilizing Physical/Chemical Properties and Models

Due to the thousands of organics which can potentially be present in society's wastes such as sewage sludge, the task of researching each organic to determine its fate in the environment is impossible. A more realistic approach would be to utilize basic physical/chemical properties of organics and soils to compare the research results assessing the environmental fate of selected organics, representative of larger groups, with the fate predicted by mathematical models.

The more important physical/chemical characteristics for assessing the potential transport, persistence, and fate of substances in sludge land applications are: (a) water solubility, (b) soil adsorption-partitioning, (c) half-life in soil, and (d) vapor pressure. Laboratory measurements can be used to obtain values for all these characteristics except soil half-life, or they may be estimated by methods such as those discussed by Lyman et al. (1982).

The fates of greatest interest for sludge organics incorporated into the soil are volatilization, degradation, plant uptake, and leaching. The persistence, or ease of degradation, and volatilization of an organic are major characteristics which will affect the time during which an organic may be "available" for plant uptake or loss by leaching to groundwater. Adsorption to soil colloids (organic and inorganic) and water solubility of an organic are also important factors which help determine this availability. When plant uptake and leaching are not significant for an organic, potential for transfer back to man is reduced. Likewise, when an organic is completely degraded in soils, additional pathways (discussed earlier in this section) for transferring a sludge organic to humans are eliminated.

Examples of using benchmark properties in mathematical models or for estimating the behavior of organics applied to soils include Chang and Page (1984), Jury et al. (1983, 1984a,b), and Wilson et al. (1981). Based on calculations using the soil adsorption coefficient, water-air partition coefficient, and octanol-water partition coefficient, Chang and Page (1984) compared the environmental fate and transport in soils of several pesticides with several trace organics. Their conclusion regarding the addition of wastewater organics to soils was that their environmental impact was not expected to be very significant.

Using a simple mathematical model based on water solubility of an organic chemical and the organic carbon content of the soil, Wilson et al. (1981) were able to predict the retardation factors for 13 organic pollutants within a factor of three. They found retardation by soil with respect to water movement generally increased with decreasing water solubility.

Jury et al. (1983) developed a more complicated mathematical model for describing transport and loss of soil-applied organic chemicals. This screening model uses benchmark properties (organic C partition coefficient, vapor pressure, solubility, half-life) to determine the relative convective mobility, diffusive mobility, volatility, and persistence (Jury et al., 1984a,b). When this model was tested on published experimental data for volatilization, leaching, and persistence, experimental results and those predicted by the model agreed reasonably well (Jury et al., 1984a,b).

Although experiments under field conditions are the only reliable way to determine the fate of an organic applied to soils with sewage sludge, the expense and time required to test the large number of organic chemicals used in society and found in sewage sludges are prohibitive. Therefore, models like those mentioned above can and should play a significant role in assessing the environmental risk of applying sludge organics to soils.

COMPARISON OF MUNICIPAL SLUDGE EXPOSURE/RISK ASSESSMENTS

Several independent evaluations have recently been published to assess the relative risk from specific organic

compounds present in municipal sludge when applied to land
(Connor, 1984; EPA, 1985; METRO, 1983; Munger, 1984; Naylor and
Loehr, 1982a,b). All of the risk assessments cited above are
published in non-peer reviewed journals. No in-depth scientific
evaluation or analysis was performed on these individual expo-
sures/risk assessments. Therefore, their results and conclusions
should be viewed with this in mind. As far as the authors are
aware, no risk assessments for sludge have appeared to date in
a peer-reviewed journal.

Risk could be defined as a measurement of the probability of
harm occurring to human health as a result of an organic chemical
being present in land-applied sludge (Munger, 1984). To the
extent possible a common concept was used to assess "acceptable
risk", i.e., the ratio of daily intake required to stay below a
risk level of 10^{-6}. If this ratio is greater than 1.0, the
resulting risk level is (numerically) greater than 10^{-6} and when
less than 1.0, the risk level is less than 10^{-6}. This ratio is
actually the inverse of the "hazard index" used in EPA environmental profiles (EPA, 1985) but was the method chosen to express
risk levels in the other risk analyses.

A risk assessment concerning the health effects of land
applying sludge was prepared by S. Munger for the Municipality
of Metropolitan Seattle (METRO, 1983). This assessment was
updated and expanded specifically for the "Municipal Wastewater
Sludge Health Effects Research Planning Workshop" held in
Cincinnati by EPA in January 1984 (Munger, 1984). That estimate
of risk was based on the quantities of soil, water, or food
obtained from a sludge application site which could be consumed
without exceeding a risk level of 10^{-5}.

The assumptions and values used by Munger (1984) for
assessing the risk of two organics, polychlorinated biphenyls
(PCBs) and benzo(a)pyrene [B(a)P], are shown in Table 46. The
concentrations of PCBs and B(a)P which would occur in the various
environmental compartments (i.e., soil, water, plant and animal
tissues) were estimated. Using these concentrations (Munger,

Table 46. Assumptions/values used for Metro analysis (Munger, 1984).

Sludge contains:	1.1 mg/kg DW PCBs (Metro sludge)
	2.6 mg/kg DW B(a)P (Metro sludge)
Application rate:	45 mt/ha for silviculture
Estimated soil concentration:	Calculated assuming even mixing in top 15 cm
Risk level used as benchmark:	10^{-5} (Values from Munger, 1984 were divided by 10 to give a 10^{-6} risk level for values in Tables 47 and 48.)
Normal daily dietary intake:	8,700 ng PCBs/day
	160 - 1,600 ng B(a)P/day
Consumption equivalent to a lifetime cancer risk of 10^{-5}:	204 ng PCBs/day
	61 ng B(a)P/day

1984) and the organic chemical consumption equivalent to 10^{-6} risk level, the quantities of various materials that could be consumed without exceeding this risk were calculated (Tables 47 and 48). The quantities for materials from sludge-amended areas could then be compared to similar materials from control (untreated) areas to evaluate any increased risk due to sludge application to forest land. The author (Munger, 1984) concluded that for these two organics, PCBs and B(a)P, any increased risk would be minimal and could be controlled by proper site management.

Naylor and Loehr (1982a,b) considered several perspectives on risk, including the human intake routes, by comparing the addition of sludge organics to soils with the application of agricultural chemicals to soil. Naylor and Loehr (1982a)

Table 47. Metro assessment of lifetime cancer risk for PCB.[*]

Environmental compartment	Estimated quantities of environmental compartments which can be consumed on a daily basis without exceeding a lifetime cancer risk of 10^{-6} based on PCB concentrations				
	Months after application				
	0	3	6	12	24
Sludge-soil (g/day)	0.8	0.8	0.9	1.0	1.0
Control soil (g/day)			≤ 2.0		
Surface water (liters/day)			≤ 2.0		
Control water (liters/day)			≤ 2.0		
Edible plants (g/day)			20-100		
Control plants (g/day)			≤ 200		
Deer fat (g/day)			2		
Control deer fat (g/day)			unknown		
Groundwater (liters/day)			≤ 2		

[*]Munger, 1985. See Table 46 for assumptions used.

Table 48. Metro assessment of lifetime cancer risk for B(a)P.[*]

Environmental compartment	Estimated quantities of environmental compartments which can be consumed on a daily basis without exceeding a lifetime cancer risk of 10^{-6} based on B(a)P concentrations				
	Months after application				
	0	3	6	12	24
Sludge-soil (g/day)	0.1	0.2	0.6	0.6-6	0.6-6
Control soil (g/day)			0.6-6		
Edible plants (g/day)		1-10	2-20	30-300	30-300
Control plants (g/day)			30-300		
Animal tissue			unknown		
Groundwater and control groundwater (liters/day)			≤ 0.6		

[*]Munger, 1985. See Table 46 for assumptions used.

begin by defining relative toxicity categories (Table 49) and then comparing the toxicities of common chemicals (Table 50), common pesticides (Table 51) and selected priority pollutant organics in 13 sludges (Table 52). In general, the agricultural chemicals are more toxic than the sludge organics. When comparing normal pesticide rates, the projected application rates for sludge organics are usually 10 fold less than for pesticides. In this perspective municipal sludge practices were judged to have no greater risk than using agricultural chemicals.

A second form of comparison was more similar to that used in the EPA Profiles for direct consumption of sludge against acute (LD$_{50}$ dose) and chronic (acceptable daily dose, or D$_T$) concerns. From the group of 24 sludge organics listed, Naylor and Loehr (1982b) selected three for further evaluation: (1) hexachloro-butadiene (HCBD), a highly toxic chemical; (2) bis-2-ethyl-hexyl phthalate, a chemical having a relatively low toxicity but present in high concentrations; and (3) 1,1,2-trichloroethane, a chemical of moderate concentration and moderate toxicity.

Table 49. Definitions for "relative toxicity" categories as used by Naylor and Loehr (1982a).

Ratings	Acute oral LD$_{50}$ mg/kg	Relative toxicity	Probable lethal oral dose of the pure chemical for a 70-kg human adult
Supertoxic	<5	6	a taste to 7 drops
Extremely toxic	5-50	5	7 drops to a teaspoon
Very toxic	50-500	4	1 teaspoon to 1 ounce
Moderately toxic	500-5,000	3	1 ounce to 1 pint (1 pound)
Slightly toxic	5,000-15,000	2	1 pint to 1 quart (2 pounds)
Practically non-toxic	>15,000	1	more than 1 quart

Table 50. Examples of chemicals commonly consumed or used and their toxicity ratings (Naylor and Loehr, 1982a).

Chemical	Acute oral LD$_{50}$ for rats,* mg/kg	Toxicity rating[†]
Sodium chloride	3,000	3
Sugar	25,800	1
Aspirin	1,000	3
Nicotine		4
Oxalic acid (present in chard, spinach, rhubarb leaves, etc.)	375	4
Caffeine	192	4
Ethyl alcohol	14,000	2
Safrole (80%) of oil of sassafras)	-	5
Gasoline, kerosene	-	3
Antifreeze	-	3
Strychnine	-	6
Cayenne pepper	-	3
Laundry bleaches	-	3-4
Aftershave lotions	-	3
Vanilla and lemon extract	-	1
Bouncing putty	-	3

*Lewis and Tatkin (1980)

†For interpretation, see Table 49

Table 51.   Toxicities and application rates for several pesticides
(Naylor and Loehr, 1982a).

Pesticide	Acute oral LD$_{50}$ for rats* mg/kg	Relative toxicity†	Recommended application rate of active ingredient to soil, ‡ lbs/acre
Methyl parathion	6	5	0.25
Parathion	2	6	0.25 to 0.38
Malathion	885	3	1.0
Diazinon	76	4	1.0
Dilox	945	3	0.5 to 1.0
2,4-D	375	4	1.0
Methoxychlor + Malathion	5,000	2	1.5 + 1.5
Diazinon	76	4	1.0 to 2.0
Sevin	250	4	1.0 to 2.0
Disyston	5	6	1.0
Dasinit	2	6	0.75 to 1.0
Dyfonate	3	6	0.75 to 1.0
Lorsban	145	4	0.75 to 1.0
Phosdrin	4	6	0.125
Maneb 80	6,750	2	up to 2.0
Sencor	2,200	3	0.38
Systox	1,700	3	0.25 to 0.5

*Lewis and Tatkin (1980)

†See Table 49 for explanation

‡New York State Coll. of Agric. & Life Sci. (1982a,b)

Table 52.   Toxicities, sludge concentrations, and projected application loadings for selected priority pollutant organics (Naylor and Loehr, 1982a).

Chemical	Acute oral LD$_{50}$ for rats* mg/kg	Toxicty rating†	No. times detected in combined sludge	Concentration in combined sludges‡				Projected application rate kg/ha, dry	
				µg/l, wet		mg/kg, dry			
				median	range	median	range	median	range
bis-2-ethylhexyl phthalate	31000	1	13	3806	157-11257	109	4.1-273	1.2	0.053-2.1
chloroethane	volatile	-	2	1259	517- 2000	19	14.5- 24	0.17	0.16-0.17
1,2-trans-dichloroethylene	volatile	-	11	744	42-54993	21	0.72-865	0.24	0.009-8.4
toluene	5000	2	12	722	54-26857	15	1.4-705	0.16	0.018-1.3
butylbenzyl phthalate	3160	3	11	577	1-17725	15	0.52-210	0.11	0.0063-1.4
2-chloronaphthalene	2078	3	1	400	400	4.7	4.7	0.03	0.03
hexachlorobutadiene	90	4	2	338	10- 675	4.3	0.52-8	0.03	0.0063-0.054
phenanthrene	700	3	12	278	34- 1565	7.4	0.89-44	0.05	0.009-0.53
carbon tetrachloride	2800	3	1	270	270	4.2	4.2	0.041	0.041
vinyl chloride	500	3	3	250	145- 3292	5.7	3-110	0.064	0.02-1.3
dibenzo (a,h) anthracene	-	-	1	250	25	13	13	0.16	0.16
1,1,2-trichloroethane	1140	3	2	222	3- 441	3.5	0.036-6.9	0.034	0.0002-0.068
anthracene	-	-	13	272	34- 1565	7.6	0.89-44	0.050	0.009-0.53
naphthalene	1780	3	9	238	23- 3100	7.5	0.9-70	0.070	0.01-0.59
ethylbenzene	3500	3	12	248	45- 2100	5.5	1.0-51	0.063	0.013-0.38
di-n-butylphthalate	1200	3	12	184	10- 1045	3.5	0.32-17	0.047	0.003-0.21
phenol	414	4	11	123	27- 4310	4.2	0.9-113	0.032	0.0011-1.5
methylene chloride	167	4	10	89	5- 1055	2.5	0.06-30	0.022	0.0004-0.97
pyrene	-	-	12	125	10- 734	2.5	0.33-18	0.024	0.004-0.22
chrysene	-	-	9	85	15- 750	2.0	0.25-13	0.022	0.0024-0.16
fluoranthene	2000	3	10	90	10- 600	1.8	0.35-7.1	0.016	0.0024-0.05
benzene	1400	3	11	16	2- 401	0.32	0.053-11.3	0.0027	0.0007-0.13
tetrachloroethylene	8100	2	11	14	1- 1601	0.38	0.024-42	0.0035	0.0002-0.54
trichloroethylene	4920	3	10	57	2- 1927	0.98	0.048-44	0.0125	0.00036-0.52

*National Academy of Sciences (1972)
†See Table 49 for interpretation
‡Feiler (1980)

Based on the maximum sludge concentrations, the rat or a cow would have to eat an amount of sludge equivalent to more than 10 times its body weight to ingest a LD$_{50}$ dose of the most toxic HCBD (Table 53). When considering a chronic exposure occurring by daily intake of sludge alone or soil treated with sludge (Table 54), a "pica" child would have to consume sludge for 41 years to consume a LD$_{50}$ dose of HCBD, the most toxic sludge organic Naylor

Table 53. Times and amounts of sludge which must be ingested by the rat or cow to reach LD_{50} doses of three sludge organics (Naylor and Loehr, 1982b).

Priority pollutant	Max.concn. in sludges mg/kg[*]	LD_{50} dose, mg/kg[*]	Toxicity rating[*]	Example animal	Typical animal wt, kg	LD_{50} dose	Amount of sludge equal to LD_{50} dose of chemical, kg	Time to consume LD_{50} dose of chemical,[†,‡] yrs
hexachloro-butadiene	8	90	4	rat cow	0.5 500	45 mg 45 g	5.6 5600	7.7 (2800 days) 6.2
bis-2-ethylhexyl phthalate	273	31000	1	rat cow	0.5 500	15.5 g 15.5 kg	57 57000	78 62
1,1,2-tri-chloroethane	6.9	1140	3	rat cow	0.5 500	570 mg 570 g	83 83000	113 91

[*]From Table 52.

[†]Daily food intake: rat = 20 g/day and cow = 25 kg/day, with sewage sludge (dry basis) intake equivalent to 10 percent by weight of total diet

[‡]Because of the length of exposure period to consume an LD_{50} dose of chemical, health effects observed are not necessarily equivalent to those observed where the dosage is ingested within more conventional LD_{50} test exposure periods of several days or less. Life expectancy of rat = 700 to 800 days (2 to 2.5 years) and of a lactating cow = 5 to 10 years.

and Loehr (1982b) considered. When considering the more logical case of a "pica" child consuming sludge treated soil rather than pure sludge, a safety factor of 45 to 450 was obtained (Table 54). Therefore, from a different perspective, 24 organics in sludge were judged to be a relatively low health hazard when sludges are land applied at agronomic rates.

Table 54. Evaluation of potential intake of three sludge organics due to sludge or soil with sludge ingested by a "pica" child or a cow (Naylor and Loehr, 1982b).

	Weight, kg	LD_{50} dose[‡]	Soil con-sump-tion g/d	Time to consume LD_{50} dose of priority pollutant, yrs		(D_I) Daily intake from soil, g	(D_T) Acceptable daily dose, g[§]	Safety factor[§] D_T/D_I
				Soil w/sludge	Sludge only			

Hexachlorobutadiene LD_{50} = 90 mg/kg
Maximum concentration of chemical in soil[*] = 0.027 mg/kg, in sludge[†] = 8 mg/kg

| Child | 20 | 1.8 g | 15 | 1 x 10⁴ | 41 | 4 x 10⁻⁷ | 1.8 x 10⁻⁵ | 45 |
| Cow | 500 | 45 g | 1500 | 3 x 10³ | 10 | 4 x 10⁻⁵ | – | |

Bis-2-ethylhexyl phthalate LD_{50} = 31000 mg/kg
Maximum concentration of chemical in soil[*] = 1.0 mg/kg, in sludge[†] = 273 mg/kg

| Child | 20 | 0.62 kg | 15 | 1 x 10⁵ | 415 | 1.5 x 10⁻⁵ | 6.2 x 10⁻³ | 415 |
| Cow | 500 | 15.5 kg | 1500 | 3 x 10⁴ | 104 | 1.5 x 10⁻³ | – | |

1,1,2-trichloroethane LD_{50} = 1140 mg/kg
Maximum concentration of chemical in soil[*] = 0.034 mg/kg, in sludge[†] = 6.9 mg/kg

| Child | 20 | 22.8 g | 15 | 1 x 10⁵ | 604 | 5.1 x 10⁻⁷ | 2.3 x 10⁻⁴ | 450 |
| Cow | 500 | 570 g | 1500 | 3 x 10⁴ | 151 | 5.1 x 10⁻⁵ | – | |

[*]Adapted from Naylor and Loehr (1982a).
[†]From Table 52.
[‡]Estimated LD_{50} dose = LD_{50} mg/kg x body wt. kg
[§]For humans, acceptable daily dose of toxic pollutants (D_T) = 10^{-5} x LD_{50} (safety factor of 10^5).

Connor (1984) used the same sludge characteristics as Naylor and Loehr (1982a,b) but an independent risk/exposure assessment (Table 55). Assuming a low (15 g) and high (139 g) amount of contaminated soil per day is ingested, safety factors were calculated based on soil concentrations expected from pesticide or sludge application. The safety margins were (1) greater for sludge organics than for common agricultural chemicals applied to soils and (2) greater than 1 for all chemicals except the polycyclic aromatic hydrocarbon (PAH) group which had a margin of safety of about 0.3 to 0.03.

Table 55.   Safety factors for ingesting soil containing pesticide and sludge organics (Connor, 1984).

	Application rate (kg/ha)	Soil concn.[†] (µg/g)	ADI (µg/day)	Safety factor Low diet	High diet
Pesticide					
2,4-D	1.1	0.5	21,000	2,800	300
Diazinon	1.1	0.5	140	19	2
Malathion	1.1	0.5	1,400	190	20
Methoxychlor	1.7	0.9	7,000	420	56
Parathion	0.29-0.44	0.15-0.22	350	110	14
Methyl parathion	0.29	0.14	70	33	4
Sludge	Concn. (mg/kg dry)				
Bis-2-ethylhexyphthalate	109	0.6	42,000	4,700	500
Toluene	15	0.08	29,500	25,000	2,700
Ethyl benzene	5.5	0.032	1,600	3,300	360
Di-n-butylphthalate	3.5	0.024	88,000	240,000	26,000
Phenol	4.2	0.016	100	420	45
Methylene chloride	2.5	0.011	25,000	150,000	16,000
Total PAH*	13	0.10	0.4	0.27	0.03
			Potency (kg d mg^{-1})	10^{-6} risk	10^{-6} risk
Hexachlorobutadiene	4.3	0.015	0.495	0.16	1.5
Carbon tetrachloride	4.2	0.020	0.083	0.36	3.3
Vinyl chloride	5.7	0.032	0.017	0.12	1.9
1,1,2-trichloroethane	3.5	0.017	0.057	0.21	1.9
Benzene	0.32	0.0014	0.052	0.016	0.14
Tetrachloroethylene	0.38	0.0018	0.040	0.015	0.14
Trichloroethylene	0.98	0.006	0.012	0.015	0.14
Total PAH	13	0.10	11.5	250	2300

Safety factors (ADI divided by daily consumption) and 10^{-6} risk calculated assuming 70 kg person ingesting equivalent of 15 g (low diet) or 139 g (high diet) of contaminated soil per day. See text for further explanation.

*PAH includes acenaphthene, fluoranthene, benzanthracene/chrysene, anthracene/phenanthrene, and pyrene. ADI calculated from WHO drinking water standard of 0.2 µg/liter and assuming consumption of 2 liters of water per day.

[†]Soil concentration calculated assuming an average sludge application rate of 15 tons/ha.

While Naylor and Loehr (1982b) did not develop a dietary scenario for PAH chemicals, they cited references which indicated that soils may contain natural PAH concentrations of 0.05-0.14 mg/kg and manure can contain 0.15 to 1.21 mg PAH/kg. The addition of PAH assumed by Connor (1984) in his risk assessment (Table 55), i.e., 15 mt/ha of sludge containing 13 mg/kg of PAH, would be equivalent to a soil concentration of 0.08 mg PAH/kg soil Therefore, Connor's analysis for PAH would suggest that any soil containing the background levels given above would have a low safety factor.

The fourth risk assessment reviewed for sludge organics is the environmental profiles and hazard indices conducted on several sludge constituents (EPA, 1985). The methodology used to assess the risk of a particular organic included the use of 12 indices to evaluate different pathways by which the sludge-applied organic could be exposed to plants, animals, or humans. The index which most lends itself to a comparison of the Naylor and Loehr (1982a,b) and Connor (1984) assessments is index 12, the index of human cancer risk from soil ingestion by a "pica" child.

Only two organics included in Naylor and Loehr's or Connor's list of organics had an index 12 value calculated: Hexachloro-butadiene (HCBD) and B(a)P. By inverting the EPA hazard index value, essentially a safety factor value can be obtained. The safety factor for HCBD was calculated as 238 by EPA (1985) and 45 by Naylor and Loehr (1982b), and for B(a)P was 0.018 by EPA (1985) and 0.27 for total PAH by Connor (1984). While these results vary, there was agreement in terms of whether risk was greater than or less than 10^{-6}.

Following a similar calculation as Naylor and Loehr (1982b) used for bis(2-ethylhexyl)phthalate and HCBD, a safety factor was also calculated for methylene chloride (MeCl) and phenanthrene. Comparing the safety factors of Connor's and Naylor and Loehr's results for 15 g/day soil ingestion show:

	Naylor & Loehr	Connor
BEHP	415	4,700
MeCl	4.6	150,000
PAH	35 (phenanthrene)	0.27 (total PAH)

Results tended to be in the same direction from 1.0 except for PAHs, although the magnitude of differences were substantial.

Differences between the three risk assessments can be attributed to differences in acceptable daily intake values used, sludge concentrations assumed, sludge rates assumed, quantities of soil assumed to be ingested, etc. The large differences obtained by these authors and cited above indicate how important it is to have correct assumptions and realistic data for calcu-lating safety factors. When the accepted daily intake values vary by 10^3 or more between assessments, large differences can be expected for the safety factors obtained.

Answers obtained using risk assessment methodology must provide realistic values to be useful. If nothing else, the methodology will indicate weaknesses in the assumptions or data used when unrealistic values are obtained with these models and help identify where further research data are needed. Overall, the general consensus of these risk assessments seems to indicate that organics applied to soil from sludge will not increase the health risk to animals and humans. However, the data base on which the previous statement is made is limited, and better risk assessment methodologies for land application of organics from sludge are urgently needed.

CONCLUSIONS

1. Because sewage sludges can theoretically contain thousands of organic compounds, wastewater treatment plants should identify the organics being discharged by users, particularly industry. This information should guide the testing of sludges for appropriate organics to determine concentration levels.

2. Available surveys measuring trace organic concentrations in sludges indicate that sewage sludges can have unusually high concentrations (i.e., a few percent, dry weight). Most organics, however, are present at concentrations less than 10 mg/kg, and about 30% of the organics tested were below detection limits. Based on their prevalence and potential loading to soils using agronomic or low sludge rates, sludge organics appear to present minimal risk.

3. Mutagenicity tests have been used to evaluate the safety of sewage sludges for land application. While such a test might provide an additional means of checking sewage sludges prior to land application, they are difficult to interpret and have not been correlated to mutagenic activity or biological toxicity of soil/sludge mixtures in the field. Therefore, their value for helping manage land application programs is uncertain at this time.

4. Major assimilative pathways for organic chemicals applied to the soil-plant system include adsorption, volatilization, degradation, leaching, and plant uptake. Many organics are strongly adsorbed to soil organic matter and/or undergo degradation, reducing the potential for plant uptake or leaching.

5. Due to the large number of organics which can be present in sewage sludges, mathematical models that utilize basic physical/chemical properties of an organic to predict the fate of sludge organics in soil appear to be a logical approach. However, limited field research with selected sludge organics (representative of organic chemical groups) is needed to validate how well these models will work.

6. In general, risk assessments suggest that most sludge organics will not increase the health risk to animals and humans, based on their relative toxicities and anticipated loadings to soil from agronomic or low sludge application rates. However, the current data base used for the risk assessments reviewed is limited, and better methodologies are needed to more completely assess health risks.

REFERENCES

American Petroleum Institute (API). 1983. Land treatment practices in the petroleum industry. Environmental Research and Technology, Inc., Concord, MA.

Ames, B. N.   1983.   Dietary carcinogens and anticarcinogens.
Science 221:1256-1264.

Ames, B. N., J. McCann, and E. Yamasaki.   1975.   Methods for
detecting carcinogens and mutagens with the Salmonella/mammalian-
microsome mutagenicity test.   Mutat. Res. 31:347-364.

Angle, J. S., and D. M. Baudler.   1984.   Persistence and
degradation of mutagens in sludge-amended soil.   J. Environ.
Qual. 13:143-146.

Babish, J. G., B. E. Johnson, and D. J. Lisk.   1983.   Mutagenicity of municipal sewage sludges of American cities.   Environ.
Sci. Technol. 17:272-277.

Baker, D. E., D. R. Bouldin, H. A. Elliott, and J. R. Miller.
1985.   Criteria and recommendations for land application of
sludges in the Northeast.   Penn. Agri. Exp. Sta. Bull. 851.

Baxter, J. C., M. Aguilar, and K. Brown.   1983.   Heavy metals
and persistent organics at a sewage sludge disposal site.
J. Environ. Qual. 12:311-316.

Beall, M. L. Jr., and R. G. Nash.   1971.   Organochlorine
insecticide residues in soybean plant tops: Root vs. vapor
sorption.   Agron. J. 63:460-464.

Borneff, J., G. Farkasdi, H. Glathe, and H. Kunte.   1973.   The
fate of polycyclic aromatic hydrocarbons in experiments using
sewage sludge-garbage composts as fertilizers.   Zbl. Bakt.
Hyg., I. Abt. Orig. B. 157:151-164.

Bowen, H. J. M.   1977.   Residence times of heavy metals in the
environment.   p. 1-19.   In T. C. Hutchinson et al. (ed.) Proc.
Intern. Conf. on Heavy Metals in the Environment, Toronto, 27-31
October 1975.   Institute for Environmental Studies, Univ. of
Toronto.   Toronto, Ontario, Canada.

Boyd, J. N., G. S. Stoewsand, J. G. Babish, J. N. Telford, and D.
J. Lisk.   1982.   Safety evaluation of vegetables cultured on
municipal sewage sludge-amended soil.   Arch. Environ. Contam.
Toxicol. 11:399-405.

Braude, G. L., C. F. Jelinek, and P. Corneliussen.   1975.   FDA's
overview of the potential health hazards associated with the land
application of municipal wastewater sludges.   pp. 214-217.   In:
Proc. Natl. Conf.   Municipal Sludge Management and Disposal,
Anaheim, CA.   18-20 August 1975.   Information Transfer, Inc.
Rockville, MD.

Brewer, W. S., A. C. Draper III, and S. S. Wey.   1980.   The
detection of dimethylnitrosamine and diethylnitrosamine in
municipal sewage sludge applied to agricultural soils.
Environ. Pollut. 1:37-43.

Brown, K. W., K. C. Donnelly, and B. Scott. 1982. The fate of mutagenic compounds when hazardous wastes are land treated. p. 383-397. In D. W. Shultz (ed.) Proc. Eighth Ann. Res. Symp. Land Disposal Hazardous Waste. EPA-600/9-82-002. Municipal Environmental Research Laboratory, Office of Research and Development, U.S. Environmental Protection Agency, Cincinnati, OH.

Brown, K. W., K. C. Donnelly, J. C. Thomas, and P. Davol. 1985. Mutagenicity of three agricultural soils. Sci. Total Environ. 41:173-186.

Burns and Roe Industrial Services Corp. 1982. Fate of priority pollutants in publicly owned treatment works. EPA-440/1-82/303. Effluent Guidelines Division, Office of Water Regulations and Standards, US Environmental Protection Agency, Washington, DC.

Chaney, R. L. 1984. Potential effects of sludge-borne heavy metals and toxic organics on soils, plants, and animals, and related regulatory guidelines. 56 pp. In: Proc. Workshop on the International Transportation, and Utilization or Disposal of Sewage Sludge, Including Recommendations, 12-15 Dec. 1983. Final Report PNSP/85-01, Pan American Health Organization, Washington, D.C.

Chang, A. C., and A. L. Page. 1984. Fate of wastewater constituents in soil and groundwater: Trace organics. p. 15-1 to 15-20. In G. S. Pettygrove and T. Asano (ed.). Irrigation with reclaimed municipal wastewater--A guidance manual. Calif. State Water Resources Control Board, Sacramento, CA.

Chou, S. F., L. W. Jacobs, D. Penner, and J. M. Tiedje. 1978. Absence of plant uptake and translocation of polybrominated biphenyls (PBBs). Environ. Health Perspect. 23:9-12.

Clevenger, T. E., D. D. Hemphill, K. Roberts, and W. A. Mullins. 1983. Chemical composition and possible mutagenicity of municipal sludges. J. Water Pollut. Control Fed. 55:1470-1475.

Connor, M. S. 1984. Monitoring sludge-amended agricultural soils. Biocycle, January/February, p. 47-51.

Dacre, J. C. 1980. Potential health hazards of toxic organic residues in sludge. pp. 85-102. In: G. Bitton et al. (ed.) Sludge-health risks of land application. Ann Arbor Sci. Pub., Inc., Ann Arbor, MI.

Davis, R. D., K. Howell, R. J. Oake, and P. Wilcox. 1984. Significance of organic contaminants in sewage sludges used on agricultural land. p. 73-79. In Proc. of Intern. Conf. on Environ. Contamination. Imperial College, London. July 1984. CEP Consultants, Edinburgh, UK.

Dean, R. B. and M. J. Suess (ed.). 1985. The risk to health of chemicals in sewage sludge applied to land. Waste Manag. Res. 3:251-278.

Dean-Raymond, D., and M. Alexander. 1976. Plant uptake and leaching of dimethylnitrosamine. Nature 262:394-396.

DeWalle, F. B., D. Kalman, D. Norman, J. Sung, and G. Plews. 1985. Determination of toxic chemicals in effluent from household septic tanks. EPA-600/2-85-050. US EPA, Cincinnati, OH.

Diercxsens, P., and J. Tarradellas. 1983. Presentation of the analytical and sampling methods and of results on organo-chlorines in soils improved with sewage sludges and compost. p. 59-68. In R. D. Davis et al. (eds.). Environmental effects of organic and inorganic contaminants in sewage sludge. Proc. of a workshop, Stevenage, UK, 25-26 May 1982. D. Reidel Pub. Co., Dordrecht, Holland.

Dressel, J. 1976a. Dependence of N-containing constituents which influence plant quality on the intensity of fertilization. Landwirtsch. Forsch. Sondern. 33:326-334.

Dressel, J. 1976b. Relationship between nitrate, nitrite, and nitrosamines in plants and soil. Qual. Plant. Plant Foods Hum. Nutr. 25:381-390.

Ellwardt, P. 1977. Variation in content of polycyclic aromatic hydrocarbons in soil and plants by using municipal waste composts in agriculture. p. 291-297. In Proc. Symposium on soil organic matter studies. Braunschweig, Germany, 1976.   IAEA-SM-211/31. Int. Atomic Energy Agency, Vienna.

Environmental Protection Agency. 1985. Summary of environmental profiles and hazard indices for constituents of municipal sludge. Office of Water Regulations and Standards, Washington, D.C. 20460. 64 pp.

Fairbanks, B. C., and G. A. O'Connor. 1984. Toxic organic behavior in sludge-amended soils. p. 80-83. Intern. Conf. on Environ. Contamination. Imperial College, London, July, 1984. CEP Consultants, Edinburgh, UK.

Farmer, W. J., K. Igue, W. F. Spencer, and J. P. Martin. 1972. Volatility of organochlorine insecticides from soil. I. Effect of concentration, temperature, air flow rate, and vapor pressure. Soil Sci. Soc. Am. Proc. 36:443-447.

Feiler, H. 1980. Fate of priority pollutants in publicly owned treatment works: Interim report. EPA-440/1-80-301. Effluent Guidelines Division, Water Regulations and Standards, Office of Water and Waste Management, U.S. Environmental Protection Agency, Washington, D.C.

Filonow, A. B., L. W. Jacobs, and M. M. Mortland. 1976. Fate of polybrominated biphenyls (PBBs) in soils. Retention of hexabromobiphenyl in four Michigan soils. J. Agric. Food Chem. 24:1201-1204.

Fox, C. J. S., D. Chisholm, and D. K. R. Stewart. 1964. Effect of consecutive treatments of aldrin and heptachlor on residues in rutabagas and carrots and on certain soil arthropods and yield. Can. J. Plant Sci. 44:149-156.

Friedman, D. 1985. Development of an organic toxicity characteristic for identification of hazardous waste. September 1985, Office of Solid Waste, U.S. EPA, Washington, D.C.

Fries, G. F. 1982. Potential polychlorinated biphenyl residues in animal products from application of contaminated sewage sludge to land. J. Environ. Qual. 11:14-20.

Fries, G. F. and G. S. Marrow. 1981. Chlorobiphenyl movement from soil to soybean plants. J. Agric. Food Chem. 29:757-759.

Furr, A. K., A. W. Lawrence, S. S. C. Tong, M. C. Grandolfo, R. A. Hofstader, C. A. Bache, W. H. Gutenmann and D. J. Lisk. 1976. Multielement and chlorinated hydrocarbon analysis of municipal sewage sludges of American cities. Environ. Sci. Technol. 10:683-687.

Goring, C. A. I., and J. W. Hamaker. 1972. Organic chemicals in the soil environment. Vol. 1. Marcel Dekker, Inc., NY.

Green, S., M. Alexander, and D. Leggett. 1981. Formation of N-nitrosodimethylamine during treatment of municipal waste water by simulated land application. J. Environ. Qual. 10:416-421.

Guenzi, W. D., J. L. Ahlrichs, G. Chesters, M. E. Bloodworth, and R. G. Nash (ed.). 1974. Pesticides in soil and water. 1st ed. Soil Sci. Soc. Amer., Madison, WI.

Hang, Y. D., J. G. Babish, C. A. Bache, and D. J. Lisk. 1983. Fate of cadmium and mutagens in municipal sludge-grown sugar beets and field corn during fermentation. J. Agric. Food Chem. 31:496-499.

Harms, H., and D. R. Sauerbeck. 1983. Toxic organic compounds in town waste materials: Their origin, concentration and turnover in waste composts, soils and plants. p. 38-51. In R. D. Davis et al. (eds.), Environmental effects of organic and inorganic contaminants in sewage sludge. Proc. of a workshop held at Stevenage, UK, 25-26 May 1982. D. Reidel Pub. Co., Dordrecht, Holland.

Hathaway, S. W. 1980. Sources of toxic compounds in household wastewater. EPA-600/2-80-128. U.S. EPA, Cincinnati, OH.

Hermanson, H. P., L. D. Anderson, and F. A. Gunther. 1970. Effects of variety and maturity of carrots upon uptake of endrin residues from soil. J. Econ. Entomol. 63:1651-1654.

Hopke, P. K., and M. J. Plewa. 1984. The evaluation of the mutagenicity of municipal sewage sludge. EPA-600/1-83-016.

Hopke, P. K., M. J. Plewa, J. B. Johnston, D. Weaver, S. G. Wood, R. A. Larson, and T. Hinesly. 1982. Multitechnique screening of Chicago municipal sewage sludge for mutagenic activity. Environ. Sci. Technol. 16:140-147.

Igue, K., W. J. Farmer, W. F. Spencer, and J. P. Martin. 1972. Volatility of organochlorine insecticides from soil. II. Effect of relative humidity and soil water content on dieldrin volatility. Soil Sci. Soc. Am. Proc. 36:447-450.

Iwata, Y., and F. A. Gunther. 1976. Translocation of the polychlorinated biphenyl Arochlor 1254 from soil into carrots under field conditions. Arch. Environ. Contam. Toxicol. 4:44-59.

Iwata, Y., F. A. Gunther, and W. E. Westlake. 1974. Uptake of a PCB (Arochlor 1254) from soil by carrots under field conditions. Bull. Environ. Contam. Toxicol. 11:523-528.

Jacobs, L. W., and M. J. Zabik. 1983. Importance of sludge-borne organic chemicals for land application programs. p. 418-426. In Proc. 6th Ann. Madison Conf. of Applied Research & Practice on Municipal & Industrial Waste. Madison, WI. 14-15 Sept. 1983. Dept. of Engineering and Applied Science, Univ. of Wisconsin, Madison, WI.

Jacobs, L. W., S. F. Chou, and J. M. Tiedje. 1976. Fate of polybrominated biphenyls (PBB's) in soils. Persistence and plant uptake. J. Agric. Food Chem. 24:1198-1201.

Jelinek, C. F., and G. L. Braude. 1977. Management of sludge use on land. J. Food Protection 41:476-480.

Jenkins, T. F., D. C. Leggett, L. V. Parker, J. L. Oliphant, C. J. Martel, B. T. Foley, and C. J. Diener. 1983. Assessment of the treatability of toxic organics by overland flow. CRREL Report 83-3, U.S. Army Corps of Engineers, Cold Regions Res. & Eng. Lab., Hanover, NH.

Jury, W. A., W. J. Farmer, and W. F. Spencer. 1984a. Behavior assessment model for trace organics in soil: II. Chemical classification and parameter sensitivity. J. Environ. Qual. 13:567-572.

Jury, W. A., W. F. Spencer, and W. J. Farmer. 1983. Behavior assessment model for trace organics in soil: I. Model description. J. Environ. Qual. 12:558-564.

Jury, W. A., W. F. Spencer, and W. J. Farmer. 1984b. Behavior assessment model for trace organics in soil: III. Application of screening model. J. Environ. Qual. 13:573-579.

Kaufman, D.D. 1983. Fate of toxic organic compounds in land-applied wastes. p. 77-151. In J. F. Parr, et al. (ed.) Land treatment of hazardous wastes. Noyes Data Corp., Park Ridge, NJ.

Kearney, P. C., M. E. Amundson, K. I. Beynon, N. Drescher, G. J. Marco, J. Miyamoto, J. R. Murphy, and J. E. Oliver. 1980a. Nitrosamines and pesticides. A special report on the occurrence of nitrosamines as terminal residues resulting from agricultural use of certain pesticides. Pure and Appl. Chem. 52:499-526.

Kearney, P. C., J. E. Oliver, A. Kontson, W. Fiddler, and J. W. Pensabene. 1980b. Plant uptake of dinitroaniline herbicide-related nitrosamines. J. Agric. Food Chem. 28:633-635.

Kowal, N. E. 1983. An overview of public health effects. p. 329-395. In A. L. Page et al. (ed.) Proc. of the workshop on utilization of municipal wastewater and sludge on land. Univ. of California, Riverside, CA.

Kowal, N. E. 1985. Health effects of land application of municipal sludge. EPA/600/1-85/015, 78 pp.

Lamparski, L. L., T. J. Nestrick, and V. A. Stenger. 1984. Presence of chlorodibenzo dioxins in a sealed 1933 sample of dried municipal sewage sludge. Chemosphere 13:361-365.

Landrigan, P. J., C. W. Heath, Jr., J. A. Liddle, and D. D. Bayse. 1978. Exposure to polychlorinated biphenyls in Bloomington, Indiana. Report EPI-77-35-2, Public Health Service-CDC-Atlanta.

Lee, C. Y., W. F. Shipe, Jr., L. W. Naylor, C. A. Bache, P. C. Wszolek, W. H. Gutenmann, and D. J. Lisk. 1980. The effect of a domestic sewage sludge amendment to soil on heavy metals, vitamins, and flavor in vegetables. Nutr. Rep. Int. 21:733-738.

Lewis, R. J., and R. L. Tatkin (eds.). 1980. Registry of toxic effects of chemical substances, 1979 edition. Nat. Inst. for Occupational Safety and Health, Washington, D.C.

Lichtenstein, E. P., G. R. Myrdal, and K. R. Schulz. 1964. Effect of formulation and mode of application of aldrin on the loss of aldrin and its epoxide from soils and their translocation into carrots. J. Econ. Entomol. 57:133-136.

Lichtenstein, E. P., G. R. Myrdal, and K. R. Schulz. 1965. Absorption of insecticidal residues from contaminated soils into five carrot varieties. J. Food Chem. 13:126-133.

Lichtenstein, E. P., and K. R. Schulz. 1965. Residues of aldrin and heptachlor in soils and their translocation into various crops. J. Agric. Food Chem. 13:57-63.

Lindsay, D. G. 1983. Effects arising from the presence of persistent organic compounds in sludge. p. 19-26. In R. D. Davis et al. (ed.) Environmental effects of organic and inorganic contaminants in sewage sludge. Proc. of a workshop held at Stevenage, UK, 25-26 May 1982. D. Reidel Pub. Co., Dordrecht, Holland.

Linne, C., and R. Martens. 1978. Examination of the risk of contamination by polycyclic aromatic hydrocarbons in the harvested crops of carrots and mushrooms after the application of composted municipal refuse (in German). Z. Pflanzenernahr. Bodenkd. 141:265-274.

Loper, J. C. 1980. Mutagenic effects of organic compounds in drinking water. Mutat. Res. 76:241-268.

Lue-Hing, C., D. T. Lordi, D. R. Zenz, J. R. Peterson, and T. B. S. Prakasam. 1985. Occurrence and fate of constituents in municipal sludge applied to land. Report 85-10. Department of Research and Development, The Metropolitan Sanitary District of Greater Chicago, Chicago, IL.

Lyman, W. J., W. F. Reehl, and D. H Rosenblatt (ed.) 1982. Handbook of chemical property estimation methods. 1st ed. McGraw-Hill, Inc., NY.

Majeti, V. A., and C. S. Clark. 1981. Health risks of organics in land application. J. Environ. Eng. Div. Proc. ASCE 107:339-357.

Mast, T. J., D. P. H. Hsieh, and J. N. Seiber. 1984. Mutagenicity and chemical characterization of organic constituents in rice straw smoke particulate matter. Environ. Sci. Technol. 18:338-348.

McConnell, E. E., G. W. Lucier, R. C. Rumbaugh, P. W. Albro, D. J. Harvan, J. R. Hass, and M. W. Harris. 1984. Dioxin in soil: Bioavailability after ingestion by rats and guinea pigs. Science 223:1077-1079.

McIntyre, A. E., and J. N. Lester. 1982. Polychlorinated biphenyl and organochlorine insecticide concentrations in forty sewage sludges in England. Environ. Pollut. (Ser. B) 3:225-230.

McIntyre, A. E., R. Perry, and J. N. Lester. 1981. Analysis of polynuclear aromatic hydrocarbons in sewage sludges. Anal. Letters 14:291-309.

METRO (Municipality of Metropolitan Seattle, Water Quality Division). 1983. Health effects of municipal wastewater sludge--A risk assessment. Municipality of Metropolitan Seattle, Seattle, WA.

Moza, P., I. Schuenert, W. Klein, and F. Korte. 1979. Studies with 2,4',5-trichlorobiphenyl-^{14}C and 2,2',4,4',6-pentachloro-biphenyl-^{14}C in carrots, sugar beets and soil. J. Agric. Food Chem. 27:1120-1124.

Moza, P., I. Weisgerber, and W. Klein. 1976. Fate of 2,2'-di-chlorobiphenyl-^{14}C in carrots, sugar beets and soil under outdoor conditions. J. Agric. Food Chem. 24:881-885.

Müller, H. 1976. Uptake of 3,4-benzopyrene by food plants from artificially enriched substrates. (In German) Z. Pflanzenernahr. Bodenkd. 139:685-695.

Mumma, R. O., D. R. Raupach, J. P. Waldman, J. H. Hotchkiss, W. H. Gutenmann, C. A. Bache, and D. J. Lisk. 1983. Analytical survey of elements and other constituents in central New York State sewage sludges. Arch. Environ. Contam. Toxicol. 12:581-587.

Mumma, R. O., D. R. Raupach, J. P. Waldman, S. S. C. Tong, M. L. Jacobs, J. G. Babish, J. H. Hotchkiss, P. C. Wszolek, W. H. Gutenmann, C. A. Bache, and D. J. Lisk. 1984. National survey of elements and other constituents in municipal sewage sludges. Arch. Environ. Contam. Toxicol. 13:75-83.

Munger, S. 1984. Health effects of organic priority pollutants in wastewater sludge--A risk assessment. In Proc. Municipal Wastewater Sludge Health Effects Research Planning Workshop, Jan. 10-12, 1984. U.S. EPA, Cincinnati, OH. p. 3-54 to 3-62.

Munger, S. 1985. Personal communication. Municipality of Metropolitan Seattle, 821 Second Ave., Seattle, WA 98104.

Nagao, M., Y. Takahashi, H. Yamanaka, and T. Sugimura. 1979. Mutagens in coffee and tea. Mutat. Res. 68:101-106.

National Research Council Committee on Animal Nutrition. 1972. Nutrient requirements of laboratory animals, No. 10. 2nd Rev. Ed. National Academy of Sciences Printing and Publishing Office, Washington, D.C.

Natural Resources Defense Council (NRDC) vs. Train. 1976. 8 ERC 2120, 2122-2129 and Appendix A, 8 ERC 2129-2130.

Naylor, L. M., and R. C. Loehr. 1982a. Priority pollutants in municipal sewage sludge. Biocycle, July/August, p. 18-22.

Naylor, L. M., and R. C. Loehr. 1982b. Priority pollutants in municipal sewage sludge. Part II. Biocycle, November/December, p. 37-42.

New York State College of Agriculture and Life Sciences. 1982a.
Cornell recommendations for field crops. Cornell University,
Ithaca, NY.

New York State College of Agriculture and Life Sciences. 1982b.
Cornell recommendations for commercial vegetable production.
Cornell University, Ithaca, NY.

Overcash, M. R. 1983. Land treatment of municipal effluent and
sludge: Specific organic compounds. p. 199-231. In A. L.
Page, et al. (ed.) Proc. of the workshop on utilization of
municipal wastewater and sludge on land. Univ. of California,
Riverside, CA.

Overcash, M. R., J. R. Weber, and W. P. Tucker. 1986. Toxic and
priority organics in municipal sludge land treatment systems.
EPA/600/2-86/010.

Pahren, H. R., J. B. Lucas, J. A. Ryan, and G. K. Dotson. 1979.
Health risks associated with land application of municipal
sludge. J. Water Pollut. Control Fed. 51:2588-2601.

Peters, M. E. 1985. Screening municipal wastewater treatment
plant sludges for suitability for land application. M.S. Thesis,
The Pennsylvania State University, University Park, PA.

Royce, C. L., J. S. Fletcher, P. G. Risser, J. C. McFarlane, and
F. E. Benenati. 1984. PHYTOTOX: A database dealing with the
effect of organic chemicals on terrestrial vascular plants.
J. Chem. Infor. Comput. Sci. 24:7-10.

Rygiewicz, P. 1986. Personal communication. U.S. EPA,
Environmental Research Lab., Corvallis, OR 97333.

Salmeen, I. T., R. A. Gorse, Jr., and W. R. Pierson. 1985.
Ames assay chromatograms of extracts of diesel exhaust particles
from heavy-duty trucks on the road and from passenger cars on a
dynamometer. Environ. Sci. Technol. 19:270-273.

Siegfried, R. 1975. The influence of refuse compost on the
3,4-benzopyrene content of carrots and cabbage (in German).
Naturwissenschaften 62:300.

Siegfried, R., and H. Müller. 1978. The contamination with
3,4-benzopyrene of root and green vegetables grown in soil with
different 3,4-benzopyrene concentrations (in German).
Landwirtsch. Forsch. 31:133-140.

Singh, D. 1983. The effect of land application of sludge on
concentration of certain sludge associated toxic chemicals in
Michigan soils and crops. Report, March 1983, Toxic Substances
Div., MI Dept. of Agric., Lansing, MI.

Strek, H. J., J. B. Weber, P. J. Shea, E. Mrozek, Jr., and M. R. Overcash. 1981. Reduction of polychlorinated biphenyl toxicity and uptake of carbon-14 activity by plants through the use of activated carbon. J. Agric. Food Chem. 29:288-293.

Suzuki, M., N. Aizawa, G. Okano, and T. Takahashi. 1977. Translocation of polychlorobiphenyls in soil into plants: A study by a method of culture of soybean sprouts. Arch. Environ. Contam. Toxicol. 5:343-352.

Tabak, H. H., S. A. Quave, C. I. Mashni, and E. F. Barth. 1981. Biodegradability studies with organic priority pollutant compounds. J. Water Pollut. Control Fed. 53:1503-1518.

Tomson, M., C. Curran, J. M. King, H. Wang, J. Dauchy, V. Gordy, and C. H. Ward. 1984. Characterization of soil disposal system leachates. EPA-600/2-84-101.

Wagner, K. H., and I. Siddiqi. 1971. The metabolism of 3,4-benzopyrene and benzo(e)acephanthrylene in summer wheat. (In German). Z. Pflazenernahr. Bodenkd. 127:211-218.

Weber, J. B., and E. Mrozek, Jr. 1979. Polychlorinated biphenyls: Phytotoxicity, absorption, and translocation by plants, and inactivation by activated charcoal. Bull. Environ. Contam. Toxicol. 23:412-417.

Weerasinghe, N. C. A., M. L. Gross, and D. J. Lisk. 1985. Polychlorinated dibenzodioxins and polychlorinated dibenzofurans in sewage sludges. Chemosphere 14:557-564.

Wilson, J. T., C. G. Enfield, W. J. Dunlap, R. L. Cosby, D. A. Foster, and L. B. Baskin. 1981. Transport and fate of selected organic pollutants in a sandy soil. J. Environ. Qual. 10:501-506.

Yoneyama, T. 1981. Detection of N-nitrosodimethylamine in soils amended with sludges. Soil Sci. Plant Nutr. 27:249-253.

ORGANIZATION, CONCLUSIONS, AND SUMMARY

The workshop group met in Las Vegas, Nevada, November 13-16, 1985, to assess the state of our knowledge on potential problems of trace elements and trace organics associated with the land application of municipal sewage sludges. The participants were divided into five separate but related workgroups. The topics of each workgroup, participants, and their affiliations are as follows:

I.   EFFECTS OF SOIL PROPERTIES ON ACCUMULATION OF TRACE ELEMENTS
     BY CROPS

     Lee E. Sommers, Chair; Colorado State University, Fort
       Collins, CO
     V. Van Volk, Oregon State University, Corvallis, OR
     Paul M. Giordano, Tennessee Valley Authority, Muscle
       Shoals, AL
     William E. Sopper, Pennsylvania State Univ., University
       Park, PA
     Robert Bastian, OMPC, U. S. EPA, Washington, D.C.

II. EFFECTS OF SLUDGE PROPERTIES ON ACCUMULATION OF TRACE
     ELEMENTS BY CROPS

     Richard B. Corey, Chair; University of Wisconsin,
       Madison, WI
     Larry D. King, North Carolina State University,
       Raleigh, NC
     Cecil Lue-Hing, Metropolitan Sanitary District of Greater
       Chicago, Chicago, IL
     Delvin S. Fanning, University of Maryland, College
       Park, MD
     Jimmy J. Street, University of Florida, Gainesville, FL
     John M. Walker, OMPC, U.S. EPA, Washington, D.C.

III. EFFECTS OF LONG-TERM SLUDGE APPLICATIONS ON ACCUMULATION
     OF TRACE ELEMENTS BY CROPS

     Andrew C. Chang, Chair; University of California,
       Riverside, CA
     Thomas D. Hinesly, University of Illinois, Urbana, IL

145

Thomas E. Bates, University of Guelph, Guelph, Ontario, Canada

Harvey E. Doner, University of California, Berkeley, CA

Robert H. Dowdy, USDA-ARS, University of Minnesota, St. Paul, MN

James A. Ryan, WERL, U.S. EPA, Cincinnati, OH

IV.   TRANSFER OF SLUDGE-APPLIED TRACE ELEMENTS TO THE FOOD CHAIN

Rufus L. Chaney, Chair; USDA-ARS, Beltsville, MD

James E. Smith, Jr., CERI, U.S. EPA, Cincinnati, OH

Dale E. Baker, Pennsylvania State University, University Park, PA

Randall Bruins, ECAO, U.S. EPA, Cincinnati, OH

Dale W. Cole, College of Forestry, Washington State University, Seattle, WA

V.   EFFECTS OF TRACE ORGANICS IN SEWAGE SLUDGES ON SOIL-PLANT SYSTEMS AND ASSESSING THEIR RISK TO HUMANS

Lee W. Jacobs, Chair; Michigan State University, East Lansing, MI

George A. O'Connor, New Mexico State University, Las Cruces, NM

Michael A. Overcash, North Carolina State University, Raleigh, NC

Matthew J. Zabik, Michigan State University, East Lansing, MI

Paul Rygiewicz, U.S. EPA, Corvallis, OR

Peter Machno, METRO, Seattle, WA

Ahmed A. Elseewi, Southern California Edison Company, Rosemead, CA

Sydney Munger, METRO, Seattle, WA

Each workgroup started out by reviewing the existing data base and prepared a working draft in their individual subject matter areas. The salient features of each group's findings were presented at plenary sessions attended by the entire workshop. Participants then were afforded an opportunity to provide inputs into workgroups other than the one to which they were assigned.

Following the workshop, chairs solicited participants for additional data or comments they wished to incorporate into the report. A revised draft was then prepared. These revised workgroup drafts were, in turn, reviewed by the workshop coordinators and the chairs, and recorders of other workgroups. Following this revision, the workshop coordinators, workgroup chairs and recorders met, and finalized the report. Findings of the workshop are summarized as follows:

I. Effects of Soil Properties on Accumulation of Trace Elements by Crops

- Although greenhouse pot studies may be useful to examine mechanisms and to establish relative response curves, the concentrations of trace elements in a particular crop are greater when the crop is grown in pots with sludge treated soils than when it is grown under comparable conditions in the field.

- Experiments which employ either trace element salts or sludges spiked with trace element salts do not simulate trace element uptake by crops grown on sludge-amended soils. Therefore, results of such studies do not provide a reliable basis for establishment of criteria, guidelines and regulations to control trace element concentrations of crops grown on sludge amended soils.

- Concentrations of trace elements in crops grown on sludge-amended soils vary with soil conditions such as the content of iron and aluminum oxides and soil pH. Iron and aluminum oxides in soils, sludges, and sludge-amended soils may reduce solubilities of trace elements and, in turn, their plant availabilities. In general, trace element uptake by crops (except Mo and Se) decreases with increasing soil pH.

- The pH measurement of a soil depends upon the method used to prepare the soil suspension. Suspensions of 1:1 soil:water or soil:$0.01 \underline{M}$ $CaCl_2$ have been used for measuring the pH of soils and/or sludge-amended soils. However, the $0.01 \underline{M}$ $CaCl_2$ method is preferred because it compensates for soluble salt contents in the soil-sludge mixture. Soil pH's in $0.01 \underline{M}$ $CaCl_2$ are generally lower than those measured in water and regulations based on soil pH should specify the method to be used.

- Sewage sludge additions have been effective in correcting trace element deficiencies (e.g., iron, copper and zinc) of crops, particularly those grown on calcareous soils.

- Trace metal uptake by crops grown on sludge-amended soil is not directly related to the soil's cation exchange capacity or texture. Available research data do not support the continued use of cation exchange capacity or soil texture alone to determine maximum available trace metal loadings.

II. Effects of Sludge Properties on Accumulation of Trace Elements by Crops

- Trace elements in raw sewage are associated primarily with suspended solids, and they remain as suspended solids in the sludge following wastewater treatment.

- Over the past decade, concentrations of trace elements in many publicly-operated treatment works (POTW) sludges have decreased markedly as a result of implementing industrial waste pretreatment, and this trend is expected to continue.

- During sewage treatment, addition of materials containing Fe, Al or lime reduces solubilities of metals in sludges.

- A variety of factors determine equilibrium trace element solubility in sludges, particularly the presence of trace-element precipitates (relatively pure compounds or coprecipitated with Fe, Al, or Ca precipitates), the strength of bonding to organic and mineral adsorption sites, the proportion of potential adsorbing sites filled, and the presence of dissolved ligands capable of complexing the trace elements.

- If, within the pH range normally found in soils of a given region, a sludge maintains the availability of a trace element below the level that causes phytotoxicity or potentially harmful accumulation of that element in plants, there is no need to limit land application of that sludge because of that element.

- If, within the pH range normally found in soils of a given region, a sludge maintains the availability of a trace element above the level that causes phytotoxicity or potentially harmful accumulation of that element in plants, loading limits should be established based on characteristics of the sludge and of the soil to which it is applied that interact to control the availability of that element.

- Development of methods for measuring trace-element desorption characteristics of sludges and adsorption characteristics of soils (particularly for Cd, Zn, Ni and Cu) should be given high priority.

- Immediately following land application all sludges will undergo changes which will affect trace element solubility and plant uptake. This effect is a function of sludge treatment prior to land application. Most research indicates that plant availability of sludge-derived metals stays the same or decreases with time following their land application.

III. Effects of Long-Term Sludge Applications on Accumulation of Trace Elements by Crops

- Application of Cd and Zn to soils from municipal sewage sludge will cause the Cd and Zn concentrations of crops grown on these soils to exceed those of untreated controls. When the sludge is applied at rates to satisfy the N requirement of the crop grown, the Cd and Zn contents of plant tissue remain low and at nearly constant levels with successive sludge applications.

- In sludge-treated soils maintained at pH >6.0, Cu and Ni contents of vegetative tissue may become slightly elevated. Phytotoxicity from sludge-applied Cu and Ni, however, has rarely been reported.

- Available data suggest that after four or more years following sludge application, the trace element concentration of the affected vegetative tissue would be determined by the total amounts of trace elements in the soil and would not be affected by the frequency of sludge application (e.g., single addition vs. multiple applications).

- Plant availability of sludge-borne metals is highest during the first year sludge is applied. Using the first-year response curve generated by a large single sludge addition will overestimate metal accumulation in vegetative tissue from plants grown in well stabilized sludge/soil systems.

- Field data indicate that trace element concentration in vegetative tissue will not rise after the termination of sludge applications if chemical conditions of the soil remain constant. Cadmium and Zn levels of plants grown in soils which were no longer receiving sludges either remained at the pretermination level or decreased with time.

IV. Transfer of Sludge-Applied Trace Elements to the Food Chain

- Contents of some trace elements in edible crop tissues can be increased when sewage sludges rich in these elements are applied to soils, especially to soils that are highly acidic (Cd, Zn, Ni) or alkaline (Mo). Under conditions which allow the concentration of a trace element in crops to increase substantially (responsive conditions), the relative increases in element concentration among crop species are sufficiently consistent to be used to generate input data for modeling the dietary exposure of the element. The relative increase of trace element concentration among crops may vary when the results are extrapolated from soils with average to soils with high organic matter contents, or from acidic to calcareous soils. High organic matter and high soil pH (except for Mo and Se) both reduce element uptake and would not increase risk above that determined from the risk assessment based on conditions of maximum intake.

- Relatively high and low Cd-accumulating crop types (lettuce vs. cabbage; carrot vs. beet) within a food group should be accounted for when using the FDA food groups to model the dietary intake of Cd.

- Representative food intake from birth to age 50 should be used to calculate daily Cd ingestion and not the maximum daily intake.

- Increased Cd ingestion from consumption of crops grown on sludge-amended soils can be expressed in terms of their Cd uptake relative to a reference crop (e.g., lettuce).

- Models developed to predict Cd retention by humans should consider not only Cd content of the diet but also other constituents in the diet (e.g., Fe, Zn) that affect Cd retention.

- The highest exposure to sludge-applied Cd would result from ingestion of a substantial fraction of the daily diet of foods grown in a strongly acidic vegetable garden for many years.

- Crop cultivars differ in their Cd uptake. However, in determining dietary Cd intake, these differences are less important than differences caused by crop species and soil and sludge characteristics.

- Surface application of sludge without soil-incorporation presents a greater potential risk to humans, livestock and wildlife due to possible direct ingestion of sludge-borne trace elements. The bioavailability of a trace element in ingested sludge is strongly influenced by the concentration of the element, the presence of other inter-acting elements, and the sludge redox potential. Livestock showed no harmful effects when grazing on pastures treated with sewage sludge containing median trace element concentrations.

V. Effects of Trace Organics in Sewage Sludges on Soil-Plant Systems and Assessing Their Risk to Humans

- Sewage sludges could contain thousands of trace organics. Organics discharged by major contributors to wastewater treatment plants should be identified to help select compounds for analysis in sewage sludge.

- Although some industrially derived organic compounds can be present in sewage sludge at relatively high concentrations (i.e., a few percent dry weight) most detected compounds are present at concentrations less than 10 mg/kg, dry weight.

* Results of bioassays of sludges for their mutagenic
  activity are difficult to interpret. Information obtained
  from these tests is not presently adequate to predict
  adverse environmental impacts associated with land application of sludge.

* Organic chemicals applied to soil may undergo adsorption,
  volatilization, degradation, leaching, and plant uptake.
  Many organics are strongly adsorbed to organic matter
  and/or undergo degradation, thus reducing the potential
  for plant uptake or leaching.

* Because experimental data are not always available for
  organics found in sludges, use of mathematical models
  based on physical/chemical properties of representative
  organic compounds is a logical approach to predict the
  fate of similar sludge-derived organics in soils. Field
  research with selected sludge organics, which are representative of organic chemical groups, is needed to
  calibrate and validate these models.

* No adverse effects on the growth of crops have been
  observed when sludges containing these organics are
  applied to soil at fertilizer rates for nitrogen or lower.

Dale E. Baker
Dept. of Agronomy
Pennsylvania State Univ.
119 Tyson Building
University Park, PA   16802
(814) 865-1221

Robert Bastian
U.S. EPA (WH-595)
401 M Street SW
Washington, DC 20460
(202) 382-7378

Thomas E. Bates
Dept. of Land Resource Sci.
University of Guelph
Guelph, ONT Canada N1G 2W1
(519) 824-4120 ext. 2452

Randall Bruins
ECAO
U.S. EPA
26 St. Clair Street
Cincinnati, OH 45268
(513) 569-7539/FTS 684-7539

Rufus L. Chaney
USDA-ARS
B008, BARC West
Beltsville, MD   20705
(301) 344-3324

Andrew C. Chang
Soil & Environ. Science
University of California
Riverside, CA 92521
(787-5325

Dale W. Cole
Department of Forestry
University of Washington
Seattle, WA 98195
(206) 545-1946

Richard B. Corey
Dept. of Soil Science
University of Wisconsin
Madison, WI 53706
(618) 263-4190

Harvey Doner
Plant & Soil Biol.
University of California
Berkeley, CA 94720
(415) 642-4148

Robert H. Dowdy
USDA-ARS
458 Borlaug Hall
University of Minnesota
1991 Upper Buford Circle
St. Paul, MN 55108
(612) 373-1444

Ahmed A. Elseewi
Research and Development
Southern California Edison
P. O. Box 800
Rosemead, CA 91770
(818) 302-3923

D. S. (Del) Fanning
Department of Agronomy
University of Maryland
College Park, MD 20742
(301) 454-3721

Paul M. Giordano
Agri. Research Branch
TVA
Muscle Shoals, AL   35660
(205) 386-2203

Thomas D. Hinesly
Department of Agronomy
University of Illinois
1102 South Goodwin Avenue
Urbana, IL   61801
(217) 333-9471

Lee W. Jacobs
Dept. of Crop & Soil Science
Michigan State University
East Lansing, MI   48824
(517) 353-7273

Larry D. King
Dept. of Soil Science
Box 7619
North Carolina State Univ.
Raleigh, NC 27695-7619
(919) 737-2645

Terry J. Logan
Department of Agronomy
Ohio State University
Columbus, OH   4310
(614) 422-2001

Cecil Lue-Hing
Director of R & D
Metro, Sanitary Dist. of Chicago
100 E. Erie Street
Chicago, IL 60611
(312) 751-5734

Peter Machno
Seattle METRO
821 2nd Avenue
Seattle, WA   98117
(206)447-6869

Elvia E. Niebla
EPA
Washington, DC
(202) 245-3036

George A. O'Connor
Crop & Soil Science Dept.
Box 3 Q
New Mexico State University
Las Cruces, NM 88003
(505) 646-2219

Michael Overcash
Dept. of Chemical Engineering
P. O. Box 7905
North Carolina State Univ.
Raleigh, NC 27695
(919) 737-2325

A. L. Page
Soil & Environ. Sci.
University of California
Riverside, CA 92521
(714) 787-3659

Alan B. Rubin
OWRS/Office of Water (WH-585)
U. S. EPA
401 M Street SW
Washington, DC 20460
(202) 245-3036

James A. Ryan
U.S. EPA
WERL
Cincinnati, OH   45268
(513) 569-7653
FTS 684-7653

Paul Rygiewicz
EPA
Environ. Research Lab
Corvallis, OR 97333
(503 757-4833

James E. Smith, Jr.
CERL
U.S. EPA
26 St. Clair Street
Cincinnati, OH 45268
(513) 569-7355

Lee Sommers
Colorado State University
Fort Collins, CO 80523
(303) 491-6517

William E. Sopper
Inst. for Res. on Land & Water
Penn State University
University Park, PA 16802
(814) 863-0291

Jimmy Street
2169 McCarty Hall
Soil Science Department
University of Florida
Gainesville, FL 32611
(904) 392-1951

V. Van Volk
Dept. of Soil Science
Oregon State University
Corvallis, OR 97330
(503) 754-2441

John M. Walker
U.S. EPA
WH-595
Washington, DC 20460
(202) 382-7283

Matthew Zabik
Pesticide Research Center
Michigan State University
East Lansing, MI 48824
(517)353-6376

INDEX

A

acceptable daily intake, 132
acceptable risk, 126
acidic sludge-amended soil, 68,80
acidic soil, 89
activity, 35
acute daily dose, 128
adsorbed metals, 32
adsorbing sites, 35
adsorption characteristics, 27
adsorption reactions, 35
adsorption sites, 45
adsorption, 1,27,117,123,133
adsorption/desorption, 36,102
adsorptive capacity, 30
adult food intake, 81,90
aeration, 5,15,121
agricultural chemicals, 128
agricultural pesticides, 102,111
agronomic rate, 1,54,112,130,133
aldrin, 112
alfalfa, 15,85,86
alkyl amines, 104
aluminum oxides, 5
aluminum, 31,37,39,45,46,56
aluminum-treated sludges, 17
Ames test, 114,115
ammoniacal fertilizers, 5
ammonium, 6
animal exposure, 7,102
animal food products, 116
animal tissue, 88,126
animals, 84,115,132,133
annual application rate, 1,58,
annual sludge application, 53,60
application rate, 15,26,35,39,46
aromatic amines, 104
arsenic, 1
asparagus, 81
assimilative pathways, 117,133
average adult dietary intake, 78,81